Advanced Vertical Super-Thin-Body FET Technology

Dinesh

TABLE OF CONTENTS

CHAPTER 1

Introduction

1.1 Introduction

In today's world, automation is widely encouraged in daily life. Automation is growing in both homes and organizations. This has led to rising demand for electronic gadgets like smartphones, laptops, smart TVs, and smartwatches. In view of this, improving the quality and performance of these devices is essential for technological progress. For this, understanding the internal structure of electronic gadgets is important. Working of these devices primarily depends on analog and digital integrated circuits (ICs) embedded inside them [1–4]. These ICs contain billions of connected transistors to perform their functions. Most of these are metal-oxide-semiconductor field-effect transistors (MOSFETs). MOSFET was invented by Mohamed M. Atalla and Dawon Kahng at Bell Labs in 1959 [5, 6]. Since then, research on MOSFETs has kept growing.

Before MOSFETs, bipolar junction transistors (BJTs) were commonly used in electronic circuits. Invention of FETs changed this by offering many advantages over BJTs, such as lower power consumption, lower noise, smaller size, easier design, and better thermal stability [7–9]. These benefits led to MOSFETs replacing BJTs in ICs, starting a major change in the microelectronics industry. Today, technology depends heavily on FET devices. MOSFETs are mainly made using CMOS technology [10–12], which combines n-type and p-type transistors to perform logic operations. CMOS is used everywhere in memory, processors, microcontrollers, sensors, RF circuits, and more [13–16]. Modern compact ICs like system-on-chip (SOC) and system-in-package (SIP) also rely on CMOS [17–19]. Device scaling shrinking transistor size while keeping performance intact has driven CMOS innovation. This idea was introduced by Gordon Moore in 1965 [20], predicting that the number of transistors on a chip would double every 24 months [21– 24], known as Moore's Law, this prediction has guided semiconductor development for decades. Smaller devices bring benefits like power reduction, more functions, better portability, and lower cost per chip. Thus, scaling has been the key approach in IC improvement, especially during 1975 - 2002. When device sizes were in the micrometer range, scaling 2D devices was relatively easy [24]. But as scaling pushed into the submicron range, the benefits of shrinking devices became harder to achieve. This was mainly due to intolerable short-channel effects (SCEs) [25–27]. For long-channel devices (>100 nm), SCEs are negligible [26]. But in short-channel devices (<100 nm), the channel length becomes comparable to the depletion regions of the source and drain, making SCEs worse [26, 27]. SCE issues include velocity saturation, channel length modulation, drain-induced barrier lowering (DIBL), increased leakage current (I_{off}), and threshold voltage roll-off [28– 30]. These problems limit the performance of traditional 2D FETs, making further scaling ineffective. To solve this, researchers began modifying MOS structures. Techniques like using SOI wafers and double-gate (DG) MOSFETs helped reduce SCEs [31– 37]. This shifted research focus toward non-planar 3D structures, leading to devices like FinFET and SOI Vertical Super-Thin Body (VSTB) FET [38– 42].

FinFET, invented by Hisamoto Digh and his team in 1989, is one of the most important 3D devices in microelectronics [43]. In a FinFET, a vertical fin forms the source, channel, and drain, with the gate wrapped around the fin. FinFETs offer advantages like low leakage, high drive current, faster speed, and better carrier mobility [44, 45]. Depending on oxide thickness layers, FinFETs can be single, double, or triple-gate devices. Triple-gate FinFETs are most common in modern technology due to better gate control [46, 47]. For scaling, the aspect ratio (AR) of fins must increase as device size shrinks to maintain current [48]. This results in extremely thin fins at advanced nodes. However, making high AR fins below 10 nm is difficult, needing more process steps and raising costs [49, 50]. These thin fins are also mechanically fragile, reducing structural stability [50]. Moreover, forming double or triple gates on such thin fins is hard [42]. Practically, FinFET AR is limited to about 3 [43], but further scaling demands much higher AR for high-performance and low-power designs. Because of these challenges, development of alternative device structures beyond FinFETs emerged. To overcome the challenges of FinFET technology, Viktor Koldiaev and Rimma Pirogova, co-founders of FinSCALE, proposed a new 3D device called the Vertical Super-Thin Body (VSTB) FET in 2014 [42]. The structural difference between a FinFET and a VSTB FET is shown in Fig. 1.1 [42]. A VSTB FET has a low-doped vertical silicon body formed on a dielectric wall. The bottom connects to a

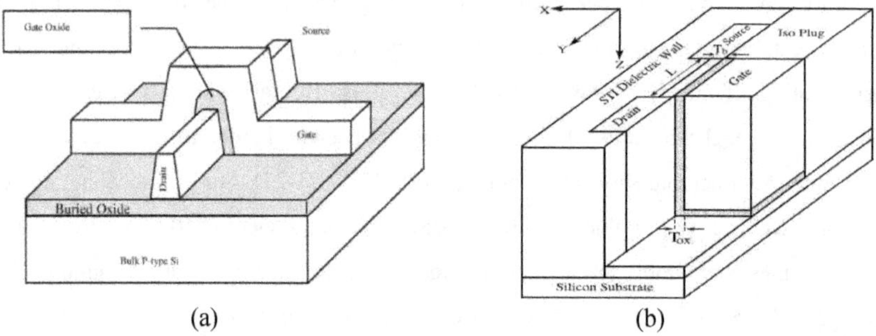

FIGURE 1.1: Structural difference between (a) FinFET, and (b) VSTB FET [43]

bulk wafer, while the top provides isolation. The channel, gate dielectric, and gate are built on the side opposite to the shallow trench isolation (STI) [50–52]. Detailed fabrication steps are also available [42]. Unlike FinFETs, the vertical body in VSTB FET is mechanically supported by STI walls [42]. This STI support allows creation of extremely thin fins, solving the mechanical stability issues of FinFETs [42]. Also, because the channel is directly next to the STI wall and sits between the source and drain, the VSTB FET works like an SOI device. This gives it SOI-like advantages. Although the VSTB FET offers several advantages due to its structural similarity with SOI-based devices, it is noteworthy that the VSTB FET is not a true SOI structure. This is because the body of the VSTB FET remains electrically connected to the substrate [42]. As a result, while the VSTB FET benefits from the mechanical support of the STI wall and achieves partial SOI-like characteristics, it does not fully inherit the electrical isolation features of a true SOI MOSFET. This body-to-substrate connection can introduce unwanted leakage paths and parasitic effects, particularly as the device scales down to nanometer dimensions. Therefore, to overcome these limitations and further enhance the performance of the VSTB FET, new structural innovations are necessary.

To address this issue, the concept of a Silicon-On-Insulator Vertical Super-Thin Body (SOI VSTB) FET has been introduced in this study. By implementing a buried oxide (BOX) layer beneath the VSTB structure, the device achieves complete electrical isolation from the substrate, combining the mechanical advantages of the STI-supported vertical body with the superior electrical properties of SOI technology [53]. By using SOI wafers, benefits such as low leakage current, low power consumption, reduced parasitic capacitance, improved resistance to latch-up, and radiation resilience can be achieved [42, 52]. The SOI VSTB FET employs a single-gate structure, which allows it to maintain high drive current even for ultra-thin channels, unlike traditional multi-gate FETs [42]. These advantages make SOI VSTB FETs a promising candidate for continued research, particularly to gain deeper insights into their internal physics, operation, and overall device performance.

1.2 The International Technology Roadmap for Semiconductors (ITRS)

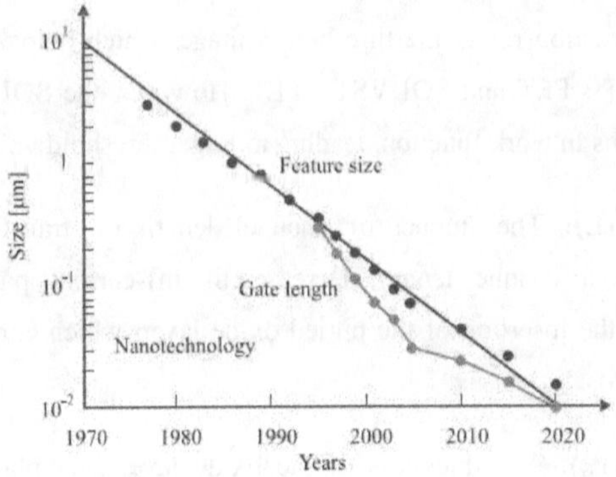

FIGURE 1.2 Feature size as a function of time [13].

A technology roadmap plays a key role by organizing specifications from device fabrication to circuit-level impact [11]. The International Technology Roadmap for Semiconductors (ITRS) has become a key reference for semiconductor research and development. It combines input from both industry and academia and confirms that Moore's law is still valid. The ITRS also highlights important research areas and provides solutions to industry challenges. Established in 1998 by the Semiconductor Industry Association (SIA), it includes members from Europe, Korea, Japan, and Taiwan [12]. The

ITRS makes three main contributions: it defines the key needs of each technology node, suggests solutions to ongoing problems, and identifies areas needing new research. It also shows how feature sizes keep shrinking over time, as seen in Fig. 1.2.

1.3 Optimization of Device Design Parameters

In the study of VSTB FETs and SOI VSTB FETs, various parameters were optimized to enhance their electrical characteristics. Here are the key optimization strategies discussed in the thesis

Doping Concentration: The doping profiles considerably influence the (I_{ON} /I_{OFF}) ratio and sub-threshold slope(SS). Lowering the doping concentration in the drain region from Doping Profile 1 (DP_1) to Doping Profile 5 (DP_5) leads to improved electrical performance, particularly in reducing off-current (I_{OFF}) without sacrificing on-current (I_{ON}) [54].

Work Function: The work function of the metal gate affects the threshold voltage. An increase in work function raises the threshold voltage, which helps in reducing off-state current in both VSTB FET and SOI VSTB FET. However, the SOI VSTB FET is more sensitive to variations in work function, leading to larger threshold voltage variations [55].

Channel Length (L): The impact of channel length on transfer characteristics is significant. Increase in channel length, decreases the off-current, particularly in the SOI VSTB FET due to the insertion of the buried oxide layer, which enhances V_{GS} influence over the channel [56].

Oxide Thickness (t_{ox}): The thickness of the oxide layer also plays a crucial role. A decrease in oxide thickness improves gate control, leading to a reduction in off-current and sub-threshold swing. This optimization is essential for maintaining the electrostatic integrity of the device [57].

Aspect Ratio (H/W): Variations in fin height (H_{fin}) and width (W_{fin}) affect the electrical characteristics. Increasing the aspect ratio generally leads to a decrease in sub-threshold swing and off-state current while increasing on-state current, which is beneficial for device performance [58].

Channel Engineering: In addition to various other techniques like single and double halo, channel materials beyond silicon are used. In the recent times, it has been shown that such technique allows scaling to some extent satisfying ITRS specifications even in 10 nm technologies [59]. Channel materials like group IV (Ge) and compound semiconductors (SiGe, GaAs, InAs, GaN, and InGaAs) are regarded as promising candidates to improve the performance of FET structure [60]. Strained SiGe channel is also used to boast the electron mobility of the device, which in turn, enhances the drive current [61].

These optimizations collectively contribute to the superior performance of the SOI VSTB FET compared to the conventional VSTB FET, particularly in terms of I_{off}, I_{on}, and sub-threshold slope.

1.4 Silicon-On-Insulator Vertical Super-Thin Body FET

The SOI VSTB FET is a cutting-edge semiconductor device developed to overcome the drawbacks of conventional MOSFETs, FinFET, and VSTB FETs, especially as device dimensions continue to shrink. In this structure (Fig. 1.3b), a thin vertical silicon channel is positioned along the side of an insulating shallow trench isolation (STI) wall, providing both mechanical support and better control over the flow of current. The gate electrode is placed along one side of this vertical silicon fin, allowing for strong electrostatic control of the channel. Beneath this silicon body, a buried oxide (BOX) layer provides electrical isolation from the substrate, helping to reduce unwanted leakage current and parasitic capacitance. This design also helps in controlling the floating body effects often seen in traditional SOI devices, improving threshold voltage stability and overall device reliability. The vertical arrangement makes it possible to place more devices in a compact area, supporting high-density integration and better power efficiency. Another important advantage of this structure is improved heat dissipation, as the STI wall helps in managing the self-heating effect, which is a common problem in highly scaled devices.

Unlike FinFETs, which require complex multi-gate designs and high aspect ratio fins, the SOI VSTB FET achieves efficient gate control with a simpler, single-gate configuration. This simplifies the fabrication process and enhances uniformity across devices. In addition, the SOI VSTB FET shows lower sub-threshold swing (SS) and reduced drain-induced barrier lowering (DIBL), making it appropriate for low-power and high-performance applications. With the growing demand for compact, energy-efficient, and high-speed electronic devices, the SOI VSTB FET stands out as a promising technology that effectively combines the strengths of SOI and vertical device architectures in a single optimized structure.

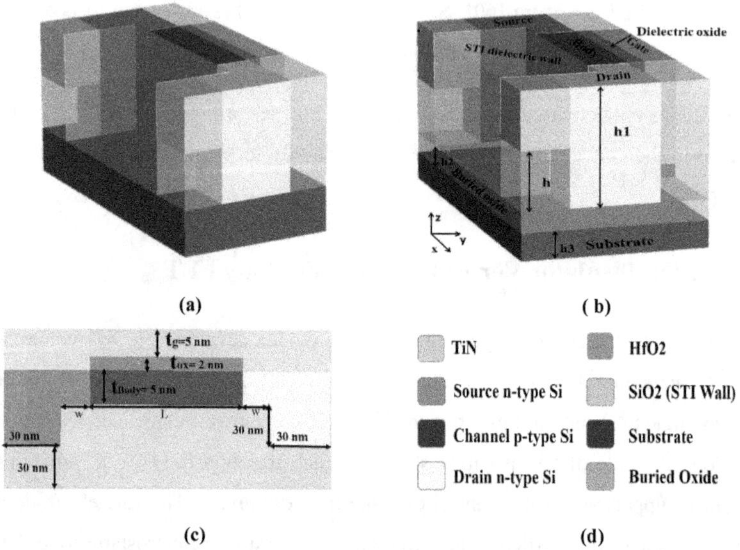

FIGURE 1.3 (a) 3D view of VSTB FET, (b) 3D view of SOI VSTB FET, (c) 2D top view, and (d) Regions name [53].

1.5 Motivation of the Thesis

The continuous push for smaller, faster, and more power-efficient electronic devices has motivated the semiconductor industry to develop advanced transistor designs beyond traditional planar MOSFETs. Although FinFETs have been successful in extending CMOS scaling, they now face serious challenges in terms of fabrication complexity, and mechanical fragility, especially at technology nodes below 10 nm. These issues have led to explore new 3D device architectures that can maintain performance while simplifying fabrication.

Among these, the VSTB FET has emerged as a promising alternative to FinFETs. The VSTB FET, with its vertically oriented thin silicon body supported by shallow trench isolation (STI) walls, offers greater mechanical stability and allows high drive current in extremely thin channels. It also provides SOI-like behavior by separating the channel from the bulk substrate, helping to reduce leakage and parasitic capacitance. However, the conventional VSTB FET is not a true SOI structure, as it still has an electrical connection to the substrate through the source region, leading to leakage and limiting its full potential. To overcome these limitations, the SOI VSTB FET is proposed in this thesis. By introducing a buried oxide (BOX) layer beneath the vertical body, the SOI VSTB FET combines the mechanical strength of the VSTB structure with the excellent electrical isolation of SOI technology. This improves short-channel control, reduces leakage, and enhances ON-state current, making the device more appropriate for low-power and high-performance applications.

Additionally, further optimization using silicon-germanium (SiGe) heterojunction channels has shown significant improvements in carrier mobility and RF characteristics, although trade-offs such as increased leakage needs to be addressed. Moreover, thermal effects like self-heating become more significant at advanced technology nodes and need careful study. Finally, with the increasing importance of electronics in space and defense sectors, radiation resilience of the device is critical, with 4H-SiC SOI VSTB FETs showing great promise for these applications.

In view of these facts, there is a need to conduct a comprehensive study on SOI VSTB FETs, focusing on their electrical, RF/analog, thermal, and radiation performance. The aim is to discover how this advanced structure can meet the future demands of next-generation CMOS and nanoelectronic technologies.

1.6 Objective of the Thesis

In light of the motivation and research gaps identified, the main objective of this thesis is to systematically explore and enhance the performance of SOI VSTB FETs for future nanoelectronic applications. The research focuses on systematically addressing several critical challenges associated with device scaling, material optimization, and reliability. Specifically, the objectives of the thesis are as follows:

1. To conduct a detailed comparative study of conventional VSTB FET and SOI VSTB FET architectures, emphasizing the impact of eliminating the body-to-substrate connection on device performance, especially in the presence of noise and leakage currents.

2. To investigate the influence of different high-k gate dielectric materials, such as HfO_2, on the electrical, RF, and analog characteristics of SOI VSTB FETs, with the goal of reducing gate leakage and improving electrostatic integrity.
3. To analyze and quantify the effects of self-heating in SOI VSTB FET structures and perform a sensitivity analysis with respect to key device parameters. This will help in identifying suitable design strategies for improving thermal reliability and ensuring stable performance at advanced technology nodes.
4. To compare and analyze the performance of similar-scale silicon and germanium VSTB FETs, incorporating architectural stress–strain effects, and to enhance carrier mobility and drive current in SOI VSTB FETs through optimized $Si_{1-x}Ge_x$ heterojunction channels, fine-tuning germanium content and channel thickness to achieve superior performance, controlled leakage, and improved device reliability.
5. To compare and evaluate Si, SiC, and 4H-SiC SOI VSTB FETs under heavy-ion and alpha-particle radiation, identifying key mechanisms and design strategies for improved radiation tolerance and reliability in harsh environments.

1.7 Organization and Outline of the Thesis

The thesis is arranged into nine chapters. The first chapter describes the overview of the thesis and the second chapter presents the literature survey related to the work carried out. The next five chapters are the work chapters, which present the methodologies and the results of the thesis objectives. Finally, overall findings and possible further scopes of the investigations are summarized in chapter nine.

Chapter 1: Introduction

This chapter describes the motivation of the research, which is driven by the need to build smaller, faster, and more energy-efficient electronic devices. Initially it discusses the shortcomings of conventional planar MOSFETs and why they have been overcome by solutions such as FinFETs, which are also encountering difficulties in the latest technology nodes. The VSTB FET is considered a viable alternative design that can mitigate the short channel effects and provide mechanical stability. Nevertheless, the SOI VSTB FET is put forth in order to enhance its electrical performance and overcome its shortcomings. This chapter sets out the thesis aim and thus the outline of the work as well as a brief description of the structure of the thesis.

Chapter 2: The Journey of MOS Scaling: A Review

In this chapter we provide a comprehensive literature review of the history of MOS transistors and the important milestones that contributed to today's state-of-the-art transistor technologies. The evolution of silicon-on-insulator, or SOI, MOSFETs is presented in detail, with particular emphasis on the emergence of vertical structures — specifically, the SOI vertical super-thin body. or VSTB, FET; the primary feature of this thesis. This chapter also discusses the details of invention and fabrication of SOI VSTB FETs. Additionally, a review of several approaches to mitigate real device effects such as noise, architectural stress, self-heating, and radiation effects is provided that serves as the basis for the research presented in this work.

Chapter 3: Electrical Characteristics of SOI VSTB FET: Comparison with Bulk VSTB FET

This chapter introduces the SOI VSTB FET structure in details and presents a comparative analysis between SOI VSTB FET and Bulk VSTB FET using TCAD simulations. Calibration of TCAD models with the experimental data is reported, so that faithful simulation results can be achieved. Key electrical performance parameters such as threshold voltage, sub-threshold swing, ON/OFF current ratio, and DIBL are analyzed. The study clearly shows how buried oxide (BOX) in SOI VSTB FET improves device isolation, leading to better short-channel behavior. This chapter establishes the electrical superiority of the proposed SOI-based architecture over the conventional bulk counterpart.

Chapter 4: RF, Analog, and Noise Performance of SOI VSTB FET.

This chapter focuses on the radio frequency (RF) and analog characteristics of SOI VSTB FETs, which are essential for high-speed and analog applications. Performance metrics like transconductance (g_m), cut-off frequency (f_T), and gain bandwidth product (GBP) are evaluated. Additionally, noise performance, including low-frequency noise effects that impact analog and RF circuits, is addressed. The performance of SOI VSTB FETs is compared with Bulk VSTB FETs, which shows their suitability for RF and analog circuit integration.

Chapter 5: Self-Heating Effects in SOI VSTB FET

The influence of self-heating effects (SHE) on SOI VSTB FET performance is examined and reported. As devices shrink in size, heat dissipation becomes a critical concern.

Temperature affects on threshold voltage, sub-threshold swing, ON/OFF current ratio, and RF performance is presented. Simulations are used to show performance degradation and suggesttions for design optimizations to minimize thermal issues, making the device more robust for advanced applications.

Chapter 6: Stress-Engineered Si/Ge SOI VSTB FETs: A TCAD Exploration of Bandgap and Performance Optimization

This chapter delves into the $Si_{1-x}Ge_x$ heterojunction channel engineering in SOI VSTB FETs to improve device electrical performance. They examine the effects of germanium content on the carrier mobility, bandgap and ON-state current and gate oxide thickness and channel dimensions. Optimizations are focused on improving both electrical and RF performance and minimizing leakage. A comprehensive comparison of silicon and germanium based devices indicates that, germanium enhances the mobility, increases the ON-state current but contribute to leakage and deteriorates sub-threshold swing. It also investigates the role of mechanical stress in enhancing performance and provides results with a balanced evaluation of performance versus power consumption.

Chapter 7: Radiation Performance of SOI VSTB FETs under Heavy Ion Radiation

This chapter focuses on the radiation resilience of SOI VSTB FETs, particularly for space and nuclear environments. Using TCAD simulations, the impact of heavy-ion and alpha-particle radiation on device behavior is studied. Comparisons are made between Si, SiC, and 4H-SiC SOI VSTB FETs, with 4H-SiC emerging as the most radiation-tolerant. Parameters like collected charge, pulsed DC current, and bipolar gain are examined in detail to evaluate the suitability of these devices for radiation-hardened applications.

Chapter 8: Radiation Performance of 14 nm SiC-4H SOI VSTB FETs under Heavy Ion Radiation

This chapter provides a comprehensive perspective on TCAD simulation results assessing the performance of SOI VSTB FETs under varying LET radiation conditions. Thus, the focus of this work is on major physical processes that take place when they are exposed to radiation, and where SOI and 4H-SiC could provide advantages that could result in stronger devices. Comparison of device performance of Si, SiC and 4H-SiC devices is made for different impact ionization locations: source, channel, and drain. On top of that, the dependence of transient current and collected charge on the angle of ion incidence is examined, which could aid in the field of radiation-hardened aerospace/nuclear electronics.

Chapter 9: Conclusion and Scope for Future Work

The final chapter sums up the key contributions of the thesis and emphasises how SOI VSTB FETs can solve many of the current limitations in transistor technology. It also outlines areas for future research, emphasizing the need for improvements in thermal design, radiation resistance, and integration into commercial semiconductor platforms.

. References

[1] T. Lin *et al.*, "Deep learning-based image analysis framework for hardware assurance of digital integrated circuits," *Microelectron. Reliab.*, vol. 123, p. 114196, 2021, doi: 10.1016/j.microrel.2021.114196.

[2] D. Chen *et al.*, "A Survey on Analog-to-Digital Converter Integrated Circuits for Miniaturized High Resolution Ultrasonic Imaging System," *Micromachines*, vol. 13, no. 1, 2022, doi: 10.3390/mi13010114.

[3] A. Pavlidis, M. M. Louerat, E. Faehn, A. Kumar, and H. G. Stratigopoulos, "SymBIST: Symmetry-Based Analog and Mixed-Signal Built-In Self-Test for Functional Safety," *IEEE Trans. Circuits Syst. I Regul. Pap.*, vol. 68, no. 6, pp. 2580–2593, 2021, doi: 10.1109/TCSI.2021.3067180.

[4] J. Yang and M. Sawan, "From Seizure Detection to Smart and Fully Embedded Seizure Prediction Engine: A Review," *IEEE Trans. Biomed. Circuits Syst.*, vol. 14, no. 5, pp. 1008–1023, 2020, doi: 10.1109/TBCAS.2020.3018465.

[5] W. M. A. 74/AA, "S. Devices and H. Dielectric," pp. 3–7, 1965.

[6] D. Kahng, "Transistors and Related Devices," no. 7, pp. 655–657, 1976.

[7] N. Camilleri, J. Costa, D. Lovelace, and D. Ngo, "Silicon MOSFETs, the microwave device technology for the 90%," *IEEE MTT-S Int. Microw. Symp. Dig.*, vol. 2, pp. 545–548, 1993, doi: 10.1109/MWSYM.1993.276880.

[8] M. Östling, R. Ghandi, and C. M. Zetterling, "SiC power devices: Present status, applications and future perspective," *Proc. Int. Symp. Power Semicond. Devices ICs*, pp. 10–15, 2011, doi: 10.1109/ISPSD.2011.5890778.

[9] C. G. Stella, M. Landani, A. Gaito, and M. Nania, "Advantage of the use of an added driver source lead in discrete Power MOSFETS," *Proc. IEEE Appl. Power Electron. Conf. Expo. (APEC)*, pp. 2574–2581, 2014, doi: 10.1109/APEC.2014.6803666.

[10] A. Zolfaghari, A. Chan, and B. Razavi, "Stacked inductors and transformers in CMOS technology," in *Phase-Locking in High-Performance Systems: From Devices to Architectures*, vol. 36, no. 4, pp. 101–109, 2003, doi: 10.1109/9780470545492.ch9.

[11] H. Iwai, "CMOS Technology Year 2010 and Beyond," *IEEE J. Solid-State Circuits*, vol. 34, no. 3, pp. 357–366, 1999, doi: 10.1109/4.748187.

[12] J. Yuan and C. Svensson, "High-speed CMOS circuit technique," *High Performance Syst. Des. Circuits Log.*, vol. 24, no. 1, pp. 362–370, 1999, doi: 10.1109/9780470544846.ch4.

[13] B. E. S. Akgul, L. N. Chakrapani, P. Korkmaz, and K. V. Palem, "Probabilistic CMOS Technology: A Survey and Future Directions," pp. 1–6, 2006.

[14] C. H. Diaz, D. D. Tang, and J. Y. Sun, "CMOS Technology for MS/RF SOC," *IEEE Trans. Electron Devices*, vol. 50, no. 3, pp. 557–566, 2003, doi: 10.1109/TED.2003.810472.

[15] B. Davari, R. H. Dennard, and G. G. Shahidi, "CMOS scaling for high performance and low power: The next ten years," *Low-Power C. Des.*, vol. 83, pp. 47–58, 1998, doi: 10.1109/9780470545058.part1.

[16] P. Jacob *et al.*, "Mitigating memory wall effects in high-clock-rate and multicore CMOS 3-D processor memory stacks," *Proc. IEEE*, vol. 97, no. 1, pp. 108–122, 2009, doi: 10.1109/JPROC.2008.2007472.

[17] M. Ali *et al.*, "A versatile SoC/SiP sensor interface for industrial applications: Design considerations," *Proc. Int. Conf. Microelectron. (ICM)*, pp. 40–43, 2019, doi: 10.1109/ICM48031.2019.9021286.

[18] I. L. Kondo *et al.*, "Heterogeneous multicore SoC with SiP for secure multimedia applications," *IEEE J. Solid-State Circuits*, vol. 44, no. 8, pp. 2251–2259, 2009, doi: 10.1109/JSSC.2009.2022219.

[19] G. Carchon *et al.*, "Thin-Film as Enabling Passive Integration Technology for RF SoC and SiP," *ISSCC*, pp. 398–399, 2005.

[20] E. Mollick, "Establishing Moore's Law," *IEEE Comput. Soc.*, pp. 62–75, 2006.

[21] P. P. Gelsinger, "Microprocessors for the New Millennium: Challenges, Opportunities, and New Frontiers," in *IEEE ISSCC*, pp. 23–26, 2001.

[22] S. Borkar, "Obeying Moore's law beyond 0.18 micron," *Proc. 13th Annu. IEEE Int. ASIC/SOC Conf.*, pp. 26–31, 2000, doi: 10.1109/ASIC.2000.880670.

[23] R. R. Tummala, "Moore's Law for Packaging to Replace Moore's Law for ICs," *Pan Pacific Microelectron. Symp.*, 2019, doi: 10.23919/PanPacific.2019.8696409.

[24] P. Gargini, "From NTRS to ITRS: The Making of an Industry," *Proc. 5th Berkeley Symp. Energy Efficient Electronic Systems (E3S)*, pp. 1–62, 2017, doi: 10.1109/E3S.2017.8246153.

[25] A. Chaudhry and M. J. Kumar, "Controlling Short-Channel Effects in Deep-Submicron SOI MOSFETs for Improved Reliability: A Review," *IEEE Trans. Device Mater. Reliab.*, vol. 4, no. 1, pp. 99–109, Mar. 2004, doi: 10.1109/TDMR.2004.824359.

[26] X. Zhou, K. Y. Lim, and W. Qian, "Threshold voltage definition and extraction for deep-submicron MOSFETs," *Solid-State Electron.*, vol. 45, no. 3, pp. 507–510, Mar. 2001, doi: 10.1016/S0038-1101(01)00035-1.

[27] A. Tsormpatzoglou, C. A. Dimitriadis, R. Clerc, G. Pananakakis, and G. Ghibaudo, "Semianalytical modeling of short-channel effects in lightly doped silicon trigate MOSFETs," *IEEE Trans. Electron Devices*, vol. 55, no. 10, pp. 2623–2631, Oct. 2008, doi: 10.1109/TED.2008.2003096.

[28] S. H. Kim et al., "High performance tri-gate extremely thin-body InAs-on-insulator MOSFETs with high short channel effect immunity and Vth tunability," *IEEE Trans. Electron Devices*, vol. 61, no. 5, pp. 1354–1360, May 2014, doi: 10.1109/TED.2014.2312546.

[29] V. Nagavarapu, R. Jhaveri, and J. C. S. Woo, "The tunnel source (PNPN) n-MOSFET: A novel high performance transistor," *IEEE Trans. Electron Devices*, vol. 55, no. 4, pp. 1013–1019, Apr. 2008, doi: 10.1109/TED.2008.916711.

[30] M. I. Khan, I. K. M. R. Rahman, and Q. D. M. Khosru, "Surface Potential-Based Analytical Modeling of Electrostatic and Transport Phenomena of GaN Nanowire Junctionless MOSFET," *IEEE Trans. Electron Devices*, vol. 67, no. 9, pp. 3568–3576, Sep. 2020, doi: 10.1109/TED.2020.3011645.

[31] J. Park, J. Colinge, and C. H. Diaz, "Pi-Gate SOI MOSFET," *IEEE Electron Device Lett.*, vol. 22, no. 8, pp. 405–406, Aug. 2001, doi: 10.1109/55.936358.

[32] Y. K. Choi et al., "Ultra-thin body SOI MOSFET for deep-sub-tenth micron era," in *Tech. Dig. Int. Electron Devices Meet.*, 1999, pp. 919–921, doi: 10.1109/IEDM.1999.824298.

[33] J. P. Colinge, FinFETs and Other Multi-Gate Transistors, *1st ed., Springer*, 2008, pp. 1–48, doi: 10.1007/978-0-387-71752-4-1.

[34] A. Sarkar, A. Kumar Das, S. De, and C. Kumar Sarkar, "Effect of gate engineering in double-gate MOSFETs for analog/RF applications," *Microelectron. J.*, vol. 43, no. 11, pp. 873–882, Nov. 2012, doi: 10.1016/j.mejo.2012.06.002.

[35] T. E. Rudenko, A. N. Nazarov, and V. S. Lysenko, "The advancement of silicon-on-insulator (SOI) devices and their basic properties," *Semicond. Phys. Quantum Electron. Optoelectron.*, vol. 23, no. 3, pp. 227–252, 2020, doi: 10.15407/spqeo23.03.227.

[36] Y. Sun, X. Yu, R. Zhang, B. Chen, and R. Cheng, "The past and future of multi-gate field-effect transistors: Process challenges and reliability issues," *J. Semicond.*, vol. 42, no. 2, 2021, doi: 10.1088/1674-4926/42/2/023102.

[37] C. Riddet, C. Alexander, A. R. Brown, S. Roy, and A. Asenov, "Simulation of ab initio quantum confinement scattering in UTB MOSFETs using three-dimensional ensemble monte carlo," *IEEE Trans. Electron Devices*, vol. 58, no. 3, pp. 600–608, Mar. 2011, doi: 10.1109/TED.2010.2095422.

[38] J. Iannacci and H. V. Poor, "Review and Perspectives of Micro/Nano Technologies as Key-Enablers of 6G," *IEEE Access*, vol. 10, pp. 55428–55458, 2022, doi: 10.1109/ACCESS.2022.3176348.

[39] T. Matsukawa et al., "Comprehensive analysis of variability sources of FinFET characteristics," in *Dig. Tech. Pap. Symp. VLSI Technol.*, 2009, pp. 118–119.

[40] R. S. Pal, S. Sharma, and S. Dasgupta, "Recent trend of FinFET devices and its challenges: A review," in *Proc. 2017 Conf. Emerg. Devices Smart Syst. (ICEDSS)*, Mar. 2017, pp. 150–154, doi: 10.1109/ICEDSS.2017.8073675.

[41] S. Inaba et al., "FinFET: The prospective multi-gate device for future SoC applications," in *Proc. 32nd Eur. Solid-State Circuits Conf. (ESSCIRC)*, 2006, pp. 50–53, doi: 10.1109/ESSCIR.2006.307528.

[42] V. Koldiaev and R. Pirogova, "Vertical super-thin body semiconductor on dielectric wall devices and methods of their fabrication," *U.S. Patent* 8,796,085 B2, Aug. 5, 2014.

[43] D. Hisamoto, "A folded-channel MOSFET for deep-sub-tenth micron era," in *IEDM Tech. Dig.*, 1998, pp. 1032–1034.

[44] D. Hisamoto et al., "FinFET—a self-aligned double-gate MOSFET scalable to 20 nm," *IEEE Trans. Electron Devices*, vol. 47, no. 12, pp. 2320–2325, Dec. 2000, doi: 10.1109/16.887014.

[45] M. Jurczak, N. Collaert, A. Veloso, T. Hoffmann, and S. Biesemans, "Review of FINFET technology," in *Proc. IEEE Int. SOI Conf.*, 2009, doi: 10.1109/SOI.2009.5318794.

[46] B. Yu et al., "FinFET scaling to 10 nm gate length," in *Tech. Dig. Int. Electron Devices Meet.*, 2002, pp. 251–254, doi: 10.1109/IEDM.2002.1175825.

[47] E. J. Nowak, I. Aller, T. Ludwig, and K. Kim, "By Doubling Up," *Boron*, no. Feb., 2004.

[48] A. Razavieh, P. Zeitzoff, and E. J. Nowak, "Challenges and Limitations of CMOS Scaling for FinFET and beyond Architectures," *IEEE Trans. Nanotechnol.*, vol. 18, pp. 999–1004, 2019, doi: 10.1109/TNANO.2019.2942456.

[49] S. Sinha, G. Yeric, V. Chandra, B. Cline, and Y. Cao, "Exploring sub-20nm FinFET design with predictive technology models," in *Proc. Des. Autom. Conf.*, 2012, pp. 283–288, doi: 10.1145/2228360.2228414.

[50] R. Pirogova, I. Wolf, and V. Koldiaev, "Investigation of novel vertical super-thin body silicon MOSFET performance advantages," in *Proc. IEEE Silicon Nanoelectron. Workshop (SNW)*, 2016, pp. 7–8, doi: 10.1109/SNW.2016.7578012.

[51] S. Roy, A. Chatterjee, D. K. Sinha, R. Pirogova, and S. Baishya, "2-D Analytical Modeling of Surface Potential and Threshold Voltage for Vertical Super-Thin Body MOSFET," *IEEE Trans. Electron Devices*, vol. 64, no. 5, pp. 2106–2112, May 2017, doi: 10.1109/TED.2017.2687465.

[52] N. P. Maity, R. Maity, S. Maity, and S. Baishya, "Comparative analysis of the quantum FinFET and trigate FinFET based on modeling and simulation," *J. Comput. Electron.*, vol. 18, no. 2, pp. 492–499, 2019, doi: 10.1007/s10825-018-01294-2.

[53] Roy Barman, K., Baishya, S. An insight to the performance of vertical super-thin body (VSTB) FET in presence of interface traps and corresponding noise and RF characteristics. *Appl. Phys. A* **125**, 865 (2019). https://doi.org/10.1007/s00339-019-3165-9

[54] K. R. Barman and S. Baishya, "Performance analysis of vertical super-thin body FET for different high-k gate dielectrics," *Appl. Phys. A*, vol. 125, p. 401, 2019, doi: 10.1007/s00339-019-2703-x.

[55] A. Kumar, N. Kar, A. Jaiswal, and A. Kar, "Impact of scaling on MOSFET performance parameters," *Int. J. Appl. Innov. Eng. Manag. (IJAIEM)*, vol. 2, no. 5, pp. 540–544, 2013.

[56] K. E. Kaharudin, F. Salehuddin, M. Z. Suhaila, and F. Waffle, "Impact of device parameters variation on FinFET analog and RF performance," *J. Eng. Sci. Technol.*, vol. 14, no. 1, pp. 2410–2430, 2019.

[57] M. Charmi, "A new analytical threshold voltage model for junctionless DG MOSFETs considering quantum confinement effect," *Chin. Phys. B*, vol. 24, no. 4, p. 047302, 2015.

[58] E. D. Kurniawan, H. Yang, C. C. Lin, and Y. C. Wu, "Reliability study of Ge-channel FinFET with different gate dielectrics," *Microelectron. Reliab.*, vol. 83, pp. 254–259, 2018.

[59] International Technology Roadmap for Semiconductors, Semicond. Ind. Assoc., San Jose, CA, USA 2013.

[60] S. K. Sinha, K. Kumar and S. Chaudhury, "Si/Ge/GaAs as channel material in nanowire-FET structures for future semiconductor devices," *2015 IEEE International Conference on Electron Devices and Solid-State Circuits (EDSSC)*, Singapore, 2015, pp. 527-530.

[61] T. Mizuno, S. Takagi, N. Sugiyama, H. Satake, A. Kurobe and A. Toriumi, "Electron and hole mobility enhancement in strained-Si MOSFET's on SiGe-on-insulator substrates fabricated by SIMOX technology," *IEEE Electron Device Letters*, vol. 21, no. 5, pp. 230-232, May 2000.

CHAPTER 2

The Journey of MOS Scaling: A Review

2.1 Introduction

The technology we see today is the result of years of continuous research and development in semiconductor devices. As the demand for better, faster, and smaller electronics grew, the semiconductor industry introduced new types of devices at different stages to meet those needs. In the early days, 2D MOSFETs were widely used in integrated circuits, and their sizes were in the micrometer range. However, when the need to fit more transistors on a single chip increased, the focus shifted toward miniaturizing devices. This need for smaller and more efficient devices led to the development of new and advanced structures over time [1–4].

This chapter provides an overview of how MOS devices evolved into the modern 3D structures used today. It begins with a brief history of MOS devices, followed by the origin of Silicon on Insulator VSTB FETs, fabrication process on the basis of VSTB FETs, and modeling of VSTB FETs. The chapter also discusses how device performance is evaluated and introduces the TCAD Sentaurus simulator, which is used for the analysis in this work. Finally, the chapter concludes with a summary of all the key points discussed.

2.2 Evolution of MOS Devices: A Brief Overview

The journey of the electronic revolution has been truly fascinating, exciting, and full of remarkable achievements. Over the years, many technological breakthroughs have played a key role in shaping the electronics industry into what it is today. Each breakthrough came at a different time and acted as a milestone, helping the world move closer to the advanced digital era we now experience. These developments have transformed the way we live, work, and communicate. This section provides a brief overview of this incredible journey, highlighting the important steps and innovations that led to the creation of modern digital technology.

2.2.1 Invention of MOSFET

Although the basic concept of a field-effect transistor (FET) was first patented by Julius Edgar Lilienfeld in 1925, it took nearly 34 years before the first working device was successfully built. In November 1959, Mohamed M. Atalla and Dawon Kahng at Bell Labs successfully fabricated the first metal-oxide-semiconductor field-effect transistor (MOSFET) [5]. Their invention was described in detail in two separate patents [6–8]. Later in 1960, Atalla proposed a method for fabricating MOS integrated circuits (ICs). He explained that, compared to bipolar junction transistors (BJTs), MOSFETs made the fabrication process simpler and allowed for better isolation of individual devices on a chip [9]. The invention of the MOSFET had a major impact on the semiconductor industry, changing its future forever. Recognizing its potential, several leading semiconductor companies such as Fairchild Semiconductor, RCA Laboratories, General Microelectronics (GMe), and IBM began developing research programs focused on MOS technology in the early 1960**s** [10–11].

In May 1964, General Microelectronics (GMe) achieved a major milestone by becoming the first company to commercialize MOSFET devices. They signed a contract with NASA, which started using MOSFETs in spacecraft and satellites [10]. By the mid-1960s, RCA Corporation had introduced MOSFETs into consumer products, such as FM radios, televisions, and amplifiers [12]. These early efforts laid the foundation for the widespread use of MOSFETs in modern electronics.

2.2.2 Scaling Trends: Moore's and Dennard's Contributions

In 1965, Gordon Moore, who was then the director of research and development at Fairchild Semiconductor, was asked by Electronics Magazine to predict how semiconductor technology would progress over the next ten years. To make this prediction, he studied how the number of components on an IC chip had been increasing each year at his company. Based on that, he predicted that the number of transistors in an IC would double every year [13]. Although Moore's prediction wasn't perfectly accurate it actually doubled nine times in that period instead of ten it was close enough to be considered highly successful [14–18]. Recognizing this small difference, Moore later revised his forecast in 1975, predicting that the number of components would now double every two years. This revised prediction is often called Moore's second law [19–23]. Moore's observation became one of the most important ideas in the history of technology. It helped companies plan their research, development, and product goals for decades. The impact of Moore's law is clear—it played a major role in the incredible progress of computing devices, making them faster, more powerful, and cheaper over time.

Along with Moore's law, another important idea, known as Dennard's scaling law, was introduced in 1974 by Robert H. Dennard and his team [24–28]. While Moore's law predicts how many transistors can fit on a chip, Dennard's law explains how the properties of transistors change as they are made smaller. It showed that as transistors shrink, their power density stays constant, while factors like capacitance, switching speed, and power consumption scale in a balanced way [24–25]. When combined, Moore's law and Dennard's law gave rise to a concept called constant field scaling or geometrical scaling, which guided the miniaturization of MOSFETs during their first major development era (1975 to 2002) [26–27]. The main goal of this scaling was to keep the electric field inside the device constant, even as the device's dimensions shrank. To achieve this, reducing the supply voltage along with the physical size of the device became a key requirement for successful scaling.

2.2.3 Short Channel Effects (SCEs) and Limitations of Dennard's Scaling

Using Dennard's scaling law along with Moore's law, engineers were able to shrink the size of 2D MOSFETs aggressively. As a result, the gate length of these devices eventually reached the sub-micron scale. However, with decreasing dimensions, the depletion regions at the source and drain junctions started to overlap with the actual length of the channel. This overlap led to several undesirable effects, known as short-channel effects (SCEs). Some of these effects include carrier velocity saturation, channel length modulation, DIBL, high leakage current (I_{off}), and threshold voltage roll-off. These issues made it difficult for the MOSFET to operate ideally as before. Another major problem was that voltage scaling, which was originally suggested by Dennard to help reduce power while shrinking device size, could no longer be used effectively because of the increase in sub-threshold leakage current [30]. Due to these challenges, around 2003, a new scaling approach called equivalent scaling was introduced. This became the main scaling strategy during the second era of MOSFET scaling (2003 to 2024) [31–33]. Unlike constant-field scaling, equivalent scaling does not rely only on shrinking physical dimensions. Instead, it uses other methods like new materials, high-k dielectrics, metal gates, and redesigning the transistor structure to improve performance while making devices smaller. Looking ahead, it was predicted that the demand to combine multiple technologies like GPS, smartphones, and tablets into small and compact devices would require reducing power consumption even further. This led to the idea of 3D power scaling, expected to guide semiconductor development from 2025 to 2040 [33–35].

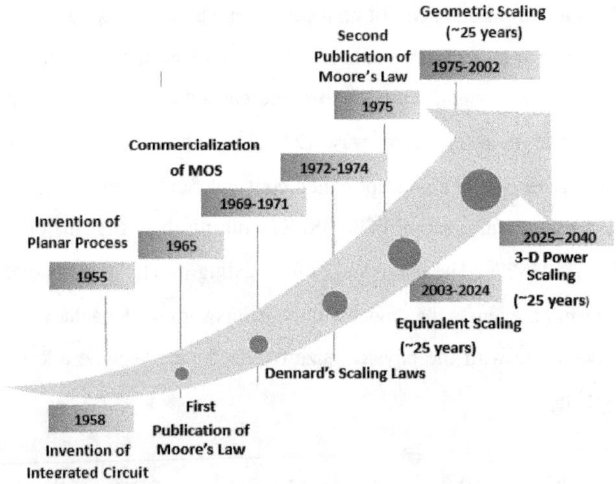

FIGURE 2.1 Timeline of Evolution of MOS Technology

In this 3D scaling approach, the main focus is on reducing power consumption, while increasing functionality. This can be achieved by using 3D transistor designs such as

FinFETs, SOI Vertical Super-Thin Body (VSTB) FETs, monolithic 3D ICs, and 3D stacked ICs [36–40]. The overall journey and evolution of MOS technology over time is illustrated in Fig. 2.1 [33].

2.3 Silicon on Insulator Vertical Super-Thin Body (VSTB) FET

This section provides a comprehensive overview of the SOI VSTB FET, starting from its conceptual development to the detailed fabrication methodologies proposed for its realization. It also covers various analytical models used to study and predict the behavior of the BULK VSTB FET. To emphasize its importance, the distinctive advantages of the SOI VSTB FET are explained in comparison with other advanced transistor technologies, showcasing its potential for addressing the challenges of modern semiconductor scaling.

2.3.1 Invention Background

The remarkable success of the microelectronics industry is the result of continuous efforts by different device technologies to solve scaling challenges at various stages of development. Among them, Silicon-On-Insulator (SOI) technology has played an important role. This technology was first proposed by C. W. Miller and P. H. Robinson in 1964. SOI devices offer several key advantages, such as lower parasitic capacitance, reduced leakage current, lower power consumption, and better resistance to latch-up [40-50]. These features make SOI technology a promising choice for modern semiconductor applications.

One key requirement for proper scaling of SOI devices is that the silicon body thickness should be about one-third of the channel length [51–56]. But when the channel length is scaled below 15 nm, making SOI wafers with extremely thin silicon bodies like 5 nm thickness becomes practically impossible [51–58]. To improve gate control, multi-gate structures like double-gate or tri-gate FETs have been used. These structures help reduce leakage current by creating more depleted charges in the off-state, restricting unwanted current flow. But even with this advantage, multi-gate FETs face a serious problem of low drive current [51–56]. This happens mainly because of carrier scattering inside the channel, caused by the multi-directional electric fields experienced by the carriers [59–61].

Among non-planar devices, FinFET is one of the most successful architectures in advanced technology nodes. In 1989, a Japanese research group led by Hisamoto Digh introduced the FinFET structure, originally called a fully depleted lean channel transistor (DELTA) [62]. Looking at its potential, improving FinFET designs to continue following Moore's law [63–67] has become a key area. FinFETs provide several benefits like low leakage current, high drive current, faster speed, and better carrier mobility [68–72]. Today, FinFETs are widely used in processors and memory by major semiconductor industries like TSMC, Intel, Samsung, Micron, Qualcomm, and others.

However, as FinFET scaling continued, new challenges started appearing. It is well-known that when we shrink the device further, height-to-width ratio (aspect ratio or AR) of the fin must be increased to maintain a high drive current. But when the channel lengths go below 10 nm, making high AR fins becomes extremely difficult and expensive [51–52]. These nanoscale 3D fins are tall and very thin, and during fabrication processes like cleaning (especially when sonication is used), they may break or get washed away [51–53]. Also, without proper mechanical support, these fins become fragile, affecting the reliability of the devices [54–55]. Additionally, making double or triple gates over such thin fins adds further complexity [51–56]. In fact, the best practical fin AR achieved in FinFETs so far is around 3 [51], whereas further scaling demands much higher AR.

To overcome these difficulties, need for a new device structure was felt. In 2014, Viktor Isakovich Koldiaev and Rimma Pirogova, co-founders of FinSCALE, proposed a new 3D device architecture called the Vertical Super-Thin Body (VSTB) FET [51–52]. As illustrated in Fig. 2.2, in this device the thin vertical fin is supported by one or more shallow trench isolation (STI) dielectric walls, which give it mechanical strength. Because of this support from STI walls, it becomes easier to manufacture high AR fins, even up to 10 [51–53]. This makes it suitable for continuing technology scaling in future nodes.

Also, because the channel is directly next to the STI wall and sits between the source and drain, the VSTB FET works like an SOI device. This gives it SOI-like advantages. Although the VSTB FET offers several advantages due to its structural similarity with SOI-based devices, it is important to note that the VSTB FET is not a true SOI structure. This is because the body of the VSTB FET remains electrically connected to the substrate [51]. As a result, while the VSTB FET benefits from the mechanical support of the STI wall and achieves partial SOI-like characteristics, it does not fully inherit the electrical isolation features of a true SOI MOS devices. This body-to-substrate connection can introduce unwanted leakage paths and parasitic effects, particularly as the device scales down to nanometer dimensions. Therefore, to overcome these limitations and further enhance the performance of the VSTB FET, new structural innovations are necessary. To address this issue, the concept of a SOI VSTB FET has been introduced in this study. By implementing a buried oxide (BOX) layer beneath the VSTB structure, the device achieves complete electrical isolation from the substrate, combining the mechanical advantages of the STI-supported vertical body with the superior electrical properties of SOI technology. Using SOI wafers, benefits such as low leakage current, low power consumption, reduced parasitic capacitance, improved resistance to latch-up, and radiation resilience can be achieved [43], [53]. The SOI VSTB FET employs a single-gate structure, which allows it to maintain high drive current even for ultra-thin channels, unlike traditional multi-gate FETs [43]. These advantages make SOI VSTB FETs a promising candidate for continued research, particularly to gain deeper insights into their internal physics, operation, and overall device performance.

2.3.2 Fabrication Flow of SOI VSTB FET: Step-by-Step Process

In the proposed SOI VSTB FET, the fabrication uses ideas from both SOI technology and the VSTB FET structure described by Viktor et al. [51]. To make the vertical super-thin body (VSTB) on an SOI wafer, either the 'gate-last' (Fig. 2.2) or 'self-aligned' (Fig. 2.3) method can be used. These methods help improve the reliability of the device and provide better mobility at lower effective oxide thickness (EOT). The process begins with an SOI wafer, which has a thin silicon layer over a buried oxide (BOX). Then, shallow trench isolation (STI) is formed to provide mechanical support for the vertical body. Next, anisotropic etching creates the vertical fin (VSTB) beside the STI wall, and the thickness of the fin depends on how much silicon remains between the trenches. After the fin is formed, a dummy isolation layer of planarized SiO_2 is deposited to prepare for source and drain regions. Vertical trenches are made in this isolation layer, and they are filled with heavily doped polysilicon to form the source and drain regions, which are activated through annealing. A recess etch then exposes the thin fin for gate formation, where a high-k dielectric and metal gate stack are deposited. Finally, iso-plugs and iso-trenches filled with insulator materials provide electrical isolation between devices. The process ends with contacts for the gate, source, and drain [51]. The fabrication process flow for the formation of the device (VSTB FET) with self-aligned approach is explained in Fig. 2.3, and the schematic diagrams of the device prototype for each step are depicted in Fig. 2.4. In the self-aligned approach, the process starts by growing a non-standard SiO_2 pad layer, followed by LPCVD Si_3N_4 as a hard mask. The ratio of the Si_3N_4 to SiO_2 is kept between 3 and 20 to control etching properly [51]. Next, recess etching is performed in the STI material to define the thickness of the VSTB. The recessed part of the STI is filled again with LPCVD Si_3N_4 to strengthen the structure.

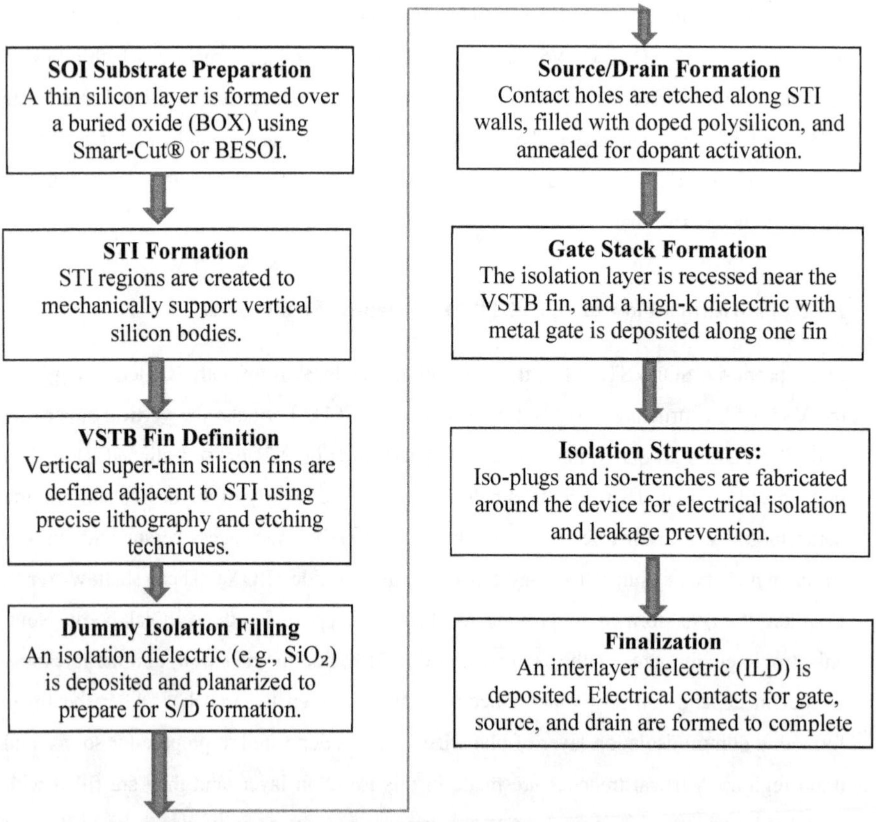

FIGURE 2.2: Proposed Stepwise Fabrication Flow of SOI Vertical Super-Thin Body (VSTB) FET Employing the 'Gate-Last' Technique [51]

Afterward, chemical mechanical polishing (CMP) is used to flatten the surface, and another SiO_2 pad oxide is deposited. Anisotropic etching is performed to expose specific regions, followed by growing a thin thermal oxide layer [51]. A dielectric cap is formed using LPCVD Si_3N_4 or Al_2O_3, and finally, the monocrystalline silicon of the SOI layer is etched to form the VSTB [51]. By combining the advantages of the SOI substrate with VSTB architecture, this fabrication process helps achieve better control of current, reduced leakage, and stronger mechanical stability, making the SOI VSTB FET suitable for future advanced electronic applications.

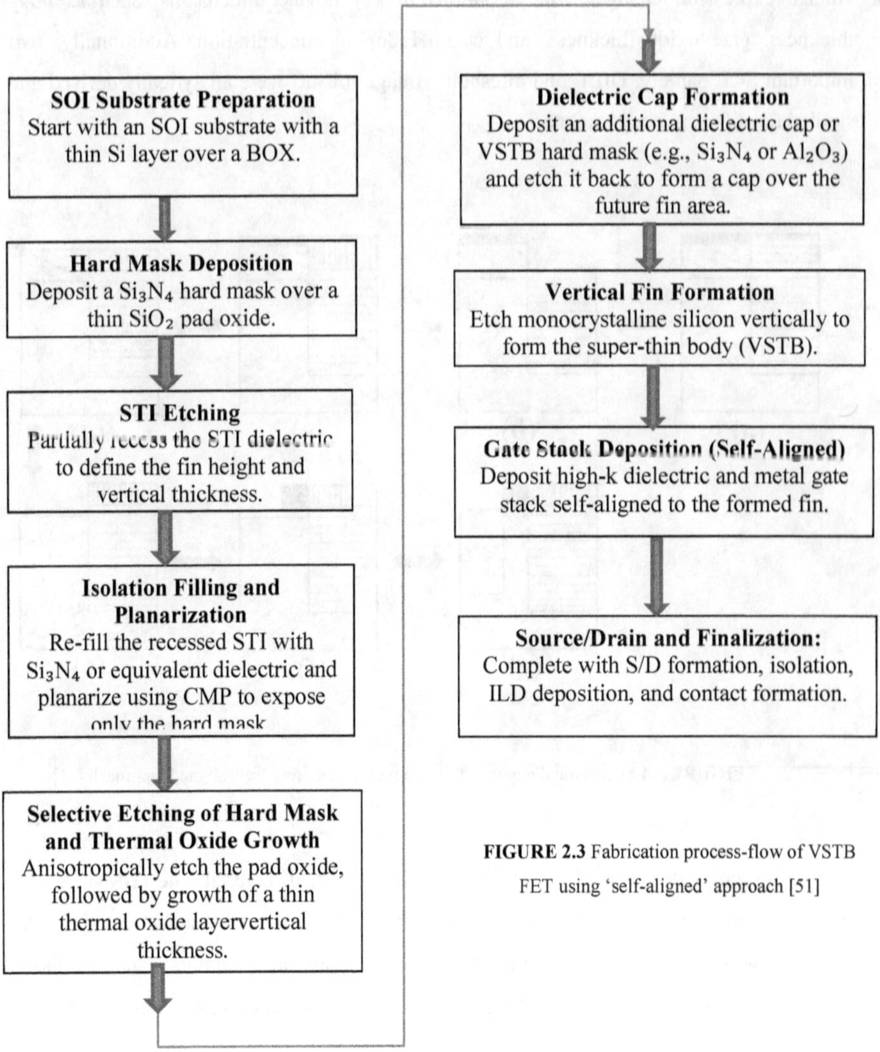

FIGURE 2.3 Fabrication process-flow of VSTB FET using 'self-aligned' approach [51]

2.3.3 2D Analytical Modeling of VSTB FETs

Saurav et al. developed a 2-D analytical model to calculate the surface potential and threshold voltage of VSTB FET by solving the 2-D Poisson's equation [55]. The parabolic approximation method was used to express the variation of surface potential both along the gate-facing surface and along the sidewall surface. The threshold voltage was then derived using the strong inversion condition, which is applied at the point where the surface potential reaches its minimum value. Their work also explored how this threshold voltage expression changes with variations in key device dimensions, such as body thickness, gate oxide thickness, and channel doping concentration. Additionally, two important SCE namely, DIBL and threshold voltage roll-off were analytically derived and studied for different body thicknesses.

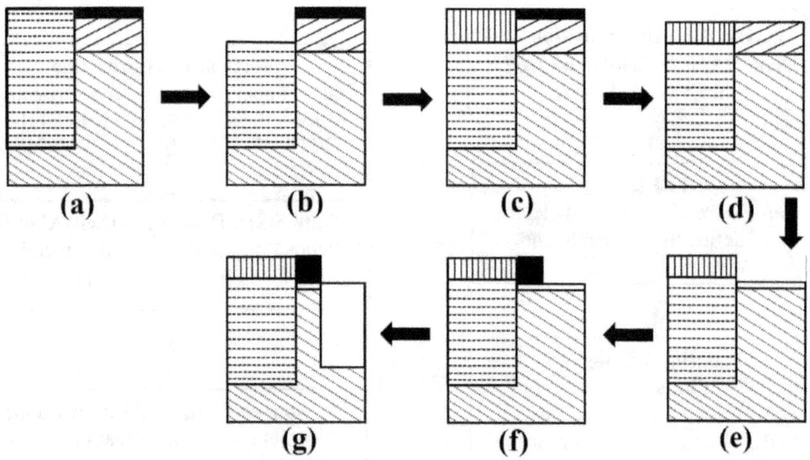

FIGURE 2.4 Schematic diagrams of device prototypes using 'self-aligned' approach [51]

Similarly, Priyanka et al. introduced a strained dual material gate (DMG) VSTB FET, presenting a 2-D analytical model for the electrostatic potential across the gate, channel, and dielectric wall interfaces [54]. They also applied the parabolic approximation technique by carefully applying suitable boundary conditions to solve for the electrostatic potential. The threshold voltage expression was formulated based on the minimum surface potential criterion, considering the contributions of both the gate and the dielectric wall potentials. Their study further examined how parameters like surface potential, electric field, threshold voltage, DIBL, and subthreshold swing (SS) behave with variations in structural parameters like body thickness, oxide thickness, and doping concentration. It was also observed that introducing strain in the channel by using an appropriate germanium (Ge) mole fraction helped improve the device performance.

2.4 Performance Metrics and Evaluation Parameters of MOS Devices

The performance of any semiconductor device needs to be thoroughly evaluated by considering multiple aspects, depending on its intended applications and usage environments. For a device to be considered reliable and efficient, its behavior must be assessed not only in terms of its static or DC performance but also through parameters related to high-frequency (RF/analog) characteristics, noise behavior, thermal effects, and reliability under different operating conditions. As technology has advanced into the nanoscale regime, this analysis has been becoming increasingly challenging. Shrinking device dimensions introduces several additional effects, including enhanced short-channel phenomena, parasitic components, variability in device parameters, and increased sensitivity to external disturbances like temperature fluctuations and radiation exposure.

These complexities mean that evaluating the overall performance of a nano-scale device is no longer limited to simple I-V characteristics but must incorporate more advanced performance metrics such as transconductance (g_m), cut-off frequency (f_T), gain-bandwidth product (GBP), sub-threshold swing (SS), DIBL, noise figure, and others. Moreover, in scaled devices, issues like self-heating effects (SHE) and work-function variability (WFV) further complicate device behavior, making their analysis critical for modern electronic design. In light of these challenges, numerous research works have reported various techniques for evaluating and optimizing the electrical and RF/analog performance of modern MOS devices; especially as the industry moves toward more compact and energy-efficient architectures.

2.4.1 Electrical and High-Frequency Performance of Modern MOS Devices

The key properties of any semiconductor device are evaluated using various electrical figures of merit (FoM), such as off-state leakage current (I_{off}), I_{on}, I_{on}/I_{off} ratio, subthreshold slope(SS), threshold voltage (V_{th}), and DIBL [73]. These parameters help determine how suitable a device is for specific applications. For low power applications, it is necessary to keep I_{off} as small as possible, while a high I_{on} is desirable for better speed and current-driving capability. According to the ITRS 2015 roadmap, for technology nodes around 4/3 nm by 2024, I_{off} should be below 100 pA/μm for low-power devices and 100 nA/μm for high-performance devices [73]. Similarly, I_{on} should exceed 2008 μA/μm for low-power and 3001 μA/μm for high-performance designs. Another critical factor is subthreshold swing (SS), which indicates how effectively the gate voltage controls the drain current. SS represents the V_{GS} required to change I_D by one order of magnitude [74].

$$SS = \frac{dv}{d(\log_{10} I_D)} \approx 2.303\emptyset_t(1 + \frac{C_{ox}}{C_{dep}}) \tag{2.1}$$

Here, V_G, I_D, C_{ox}, and C_{dep} represent the gate voltage, drain current, oxide capacitance, and depletion capacitance, respectively. Φ_t denotes the thermal voltage, which is a non-scalable quantity. For MOSFETs operating at 300 K, the lowest achievable subthreshold swing (SS) is around 60 mV/dec [74]. Another key parameter is the threshold voltage (V_{th}), which defines the gate voltage needed to start strong inversion in the channel. In modern nanoscale devices, V_{th} must be carefully balanced too low increases leakage, while too high leads to slower switching [75–78]. According to ITRS 2015, V_{th} targets are around 201 mV for low-power and 84 mV for high-performance logic applications for the 4/3 nm node by 2024 [73]. DIBL also affects how the drain voltage influences the channel's potential [73]. To ensure strong gate control, DIBL should remain minimal [79].

$$DIBL = \frac{(V_{Th,Lin} - V_{Th,Sat})}{(V_{D,Sat} + V_{D,Lin})} \tag{2.2}$$

where, $V_{Th,lin}$ ($V_{Th,Sat}$) represents the threshold voltage of the device at the drain voltage ($V_{D,lin}$ or $V_{D,Sat}$) corresponding to its linear or saturation operation. The RF and analog performance of a device is evaluated using key parameters like gate capacitance (C_{gg}), gate-drain capacitance (C_{gd}), transconductance (g_m), output conductance (g_d), intrinsic gain (g_m/g_d), cut-off frequency (f_T), and gain-bandwidth product (GBP) [80–82]. C_{gg} and C_{gd} measure the change in gate charge with a small variation in gate and drain voltage, resepctively [80]. Lower C_{gg} and C_{gd} values help improve device speed [80]. Transconductance (gm) shows how much the drain current changes with gate voltage, and higher g_m leads to better speed and drive current [80]. Output conductance (g_d) should be low because higher g_d means poor control of the channel by the gate. A higher intrinsic gain (g_m/g_d) is also desirable for better signal amplification [82]. The cut-off frequency (f_T) defines the point where the device's current gain becomes one. GBP is the product of gain and bandwidth, and higher values of f_T and GBP lead to faster circuit operation. These improvements in device speed and performance can be achieved by increasing transconductance (g_m) and reducing gate capacitances (C_{gg} and Cgd).

$$f_T = {g_m}/{2\pi C_{gg}} \tag{2.3}$$

$$GBP = {g_m}/{20\pi C_{gd}} \tag{2.4}$$

Additionally, analog performance is analyzed using parameters like transconductance generation factor (TGF = g_m/I_D), transconductance frequency product (TFP = $(g_m/I_D) \times f_T$), gain frequency product (GFP = $(g_m/g_d) \times f_T$), and gain transconductance frequency product (GTFP = $(g_m/g_d) \times (g_m/I_D) \times f_T$) [83–84].

Mohammad et al. introduced a modified SOI structure by adding a heavily doped silicon layer between the buried oxide and the source/drain, helping reduce the floating body effect [85]. Their proposed structure showed better electrical performance compared to conventional SOI MOSFETs [85]. Similarly, Nazanin et al. demonstrated that using hetero-dielectric gate oxides in FinFETs improves key parameters like off-current, on-current, and DIBL [86]. Mohapatra et al. explored how the fin height, fin width, and gate length of FinFETs affect various electrical parameters like g_m, I_{eff}, V_{th}, gain, and f_T [87]. Their study highlighted that taller fins improve drive current while thinner fins reduce short-channel and substrate effects. Tetsu et al. suggested improving propagation delay in FD SOI devices by biasing the substrate, especially with ultra-thin BOX layers [88].

Yogesh et al. investigated RF/analog linearity parameters, including higher-order coefficients and intercept points, for junctionless cylindrical surrounding gate MOSFETs with and without gate material engineering (GME) [84]. They found that GME devices perform better under interface trap conditions. Suruchi et al. further studied how temperature variations impact the RF and linearity performance of heterojunction asymmetric dopingless TFETs, showing that bandgap narrowing and mobility reduction at higher temperatures degrade overall device performance [89].

2.4.2 Impact of Interfacial Traps and Noise on Device Performance

The performance of semiconductor devices often deviates from ideal behavior due to the presence of interfacial trap charges and noise. These effects are more noticeable in nanoscale devices because smaller dimensions make them more sensitive to imperfections. Traps can form due to fixed charges during oxide growth, interface defects, or damage from etching processes [90–92]. Depending on their charge type, traps can be classified as donor or acceptor. These traps disturb the flow of current by randomly capturing and releasing carriers, which disrupts normal device operation. They can also impact carrier diffusion and recombination, resulting in diffusion noise and generation-recombination (G-R) noise [93–94]. Additionally, flicker noise often originates from contaminants within the channel [95–96].

Sheng-Chia et al. showed that interfacial traps in high-k/metal gate FinFETs significantly affect I_{off} and DIBL, especially at high trap densities [97]. They noted that if trap density (D_{it}) is below 1×10^{12} $eV^{-1}cm^{-2}$, variability in key parameters remains within 5% [97]. Jae et al. further demonstrated how gate oxide traps increase low-frequency noise in submicron FETs, with the noise worsening at higher trap densities [98]. Similarly, Jun-Sik reported that thicker silicide layers in SOI FinFETs lead to higher noise, greater parasitic capacitance, and lower mobility [99].

2.4.3 Impact of Stress and Strain on Device Performance

In modern nanoscale devices, stress and strain in the device structure can greatly affect how current flows through the channel [100–102]. Stress changes the atomic arrangement of a crystal, altering its lattice constant by increasing the space between atoms [103]–[105]. This stretching of the crystal improves carrier mobility by reducing the atomic resistance to carrier flow [106–108].

Hyeokjin et al. showed that mechanical stress from hybrid STI impacts negative bias temperature instability (NBTI) in strained Si/SiGe p-MOSFETs [105]. They also noted that tensile stress reduces the bandgap, which increases carrier concentration, but extra steps are needed to manage stress near the gate edges. Xiangchen et al. explained how fabrication processes like deposition and annealing can introduce stress in GAA Si nanowire FETs, slightly boosting drain current and transconductance [106]. Similarly, Qinglei et al. reported that applying strain to silicon and germanium nanomembranes (NMs) improves carrier mobility by modifying the crystal's lattice properties, making them promising for 3D nano-architectures [107]. Po-Yu investigated stress from gate-line-end (GLE) effects on high-k/metal gate FinFETs [108]. Their study showed that a 100 nm GLE improved drive current by about 6% in both n- and p-FinFETs [108]. They also observed that adding more fins helped reduce stress-related current degradation [108].

2.4.4 Impact of High-k Gate Dielectrics on Device Performance

In the early development of MOS technology, the Si/SiO_2 stack was the most common semiconductor-oxide combination used in devices [109–111]. This was mainly due to the excellent interface quality between silicon and silicon dioxide, making SiO_2 a preferred gate dielectric [109]. The thickness of the gate oxide controls the capacitance, which in turn governs how effectively the gate controls the channel [109–111]. To improve gate control, the oxide layer is to be made thinner, which increased gate capacitance [112]. However, as the oxide thickness approached 2 nm, tunneling current becomes a serious issue [110–112]. Use of high-k dielectrics (like HfO_2, Al_2O_3, Y_2O_3, La_2O_3) can maintain high capacitance with thicker layers, reducing leakage [113–115].

Anthony et al. reviewed the reliability of high-k materials and found that charge trapping was a major concern, leading to threshold voltage instability, while HCA and TDDB were less problematic [116]. Yung-Hsien et al. demonstrated that ZrO_2/Al_2O_3/Si stacks provide better performance than Al_2O_3/ZrO_3/Si, with reduced leakage and better reliability [117]. NH_3 plasma nitridation further improved these stacks by lowering leakage current.

Nihar et al. studied the influence of high-k dielectrics on short channel effects (SCEs) and circuit behaviour [118]. While high-k materials improved some capacitances, they worsened others, like internal fringe capacitance, which negatively impacted DIBL and noise margins [118]. To address this, using low-k sidewalls with high-k gate insulators was recommended [118]. Manoj et al. showed that high-k dielectrics increase certain parasitic capacitances in FinFETs, worsening SCEs [119]. They suggested improvements like reducing fin width, adjusting doping, and refining gate work functions to optimize device behavior at advanced nodes like 32 nm.

2.4.5 Impact of Work Function on Device Characteristics

The work function of the gate material is a critical parameter that significantly influences the electrostatic control, threshold voltage, and overall performance of nanoscale MOSFETs [120-121]. By adjusting the gate work function, device engineers can effectively modulate the threshold voltage without relying solely on channel doping, thereby reducing short-channel effects and improving device scalability [120]. De and coworkers demonstrated that optimizing the work function of the gate electrode enhances device performance at the 50 nm technology node, offering better trade-offs between drive current and leakage characteristics [120]. Similarly, Crupi et al. highlighted that variations in the interfacial layer and gate work function can also impact the low-frequency noise behavior in advanced MOSFETs with high-k dielectric stacks [121]. These findings underscore the dual role of the work function in controlling both the electrostatics and the noise characteristics of scaled CMOS devices. The ability to engineer the work function is thus indispensable for achieving desirable threshold voltage values, enhancing device reliability, and minimizing variability issues in modern CMOS technology nodes.

2.4.6 Impact of Self-Heating on Device Performance

Apart from gate bias, many other factors impact carrier movement in transistors. As carriers move along the channel, they face scattering caused by electric fields, interactions with other carriers, or collisions with lattice atoms [90, 122–125]. This scattering reduces carrier mobility and lowers the drive current [122]. In sub-micron devices, scattering increases because carriers have less space to travel without collisions [124].

According to Joule's law, flowing current generates heat, making lattice atoms vibrate [126–129]. These vibrations increase phonon interactions, further reducing carrier mobility. As the cycle continues, lattice temperature rises, causing a self-heating effect (SHE) [129, 131]. If the temperature gets too high, it can even damage the device [132–142]. Since smaller devices concentrate heat over smaller areas, SHE becomes more serious at nanoscale [143–150]. Thus, thermal management is critical to prevent device failure.

Shou-En et al. found that SHE affects device reliability, particularly bias temperature instability (BTI) and hot carrier injection (HCI) [131]. They suggested using microsecond-delay measurement systems to handle the issue [131]. Sungjae et al. studied how gate stacks and contacts help conduct heat away from the device [132]. They found that SHE is generally not severe during typical logic operation due to short heating time constants [132].

Chang et al. examined SHE effects on interconnects in FinFETs, showing that fins produce more heat than planar designs, impacting long-term heat reliability [133]. Muhammad et al. reviewed hot carrier degradation (HCD) influenced by SHE and suggested solutions like negative capacitance FETs, heat sinks, and improved electrothermal models [136].

2.4.7 Radiation Sensitivity of Modern Semiconductor Devices

Radiation effects pose significant challenges for semiconductor devices, especially as feature sizes scale down. Exposure to ionizing radiation such as protons, heavy ions, or gamma rays can result in various detrimental phenomena including total ionizing dose (TID) effects, displacement damage (DD), and single event effects (SEE). TID leads to charge buildup in the gate oxide or buried oxide layers, shifting threshold voltages and degrading device performance. Displacement damage causes defects in the silicon lattice, impacting carrier mobility and lifetime. Among these, SEEs particularly single event upsets (SEU) and single event transients (SET) are of paramount concern for nanoscale MOS devices, as they can induce erroneous switching, data corruption, or even permanent device failure [139–141].

Studies on advanced architectures such as nanowire GAA FETs, FinFETs, TFETs, and junctionless FETs (JLFETs) have revealed varying degrees of sensitivity to these radiation-induced phenomena [139–141]. For instance, the radiation response of GAA NW-FETs shows a higher susceptibility to single event transients compared to bulk CMOS due to enhanced electric fields and thinner channels [139]. Similarly, TFET-based devices, despite their potential for steep subthreshold swing and low-power applications, exhibit significant SET susceptibility due to their asymmetric doping profiles and band-to-band tunneling conduction mechanisms [141]. Heavy-ion radiation effects, often characterized by linear energy transfer (LET) studies, reveal that LET-induced transient currents increase with decreasing channel dimensions, making advanced nanoscale devices more vulnerable [142].

Comprehensive reviews highlight that radiation-induced charge collection mechanisms are strongly dependent on device geometry, doping concentration, and dielectric material selection [139-140]. For example, replacing conventional silicon channels with materials like germanium or integrating double-gate architectures can alter the sensitivity profile under irradiation [140]. Furthermore, the introduction of high-k dielectrics and scaling of oxide thickness exacerbate TID effects but can be partially mitigated by careful interface engineering and choice of gate stacks [142].

Overall, as highlighted by multiple studies [139, 141-142], achieving robust radiation resilience in next-generation devices like SOI VSTB FETs will require a combination of material engineering, architectural innovation, and accurate radiation-aware simulation frameworks. Such strategies are critical not only for terrestrial electronics but especially for aerospace, defense, and nuclear applications where radiation tolerance is mandatory.

2.5 TCAD Sentaurus Simulator

For advanced semiconductor research, especially when dealing with novel device architectures like Silicon-On-Insulator Vertical Super-Thin Body Field Effect Transistors (SOI VSTB FETs), simulation tools play a crucial role in predicting, analyzing, and optimizing device behavior before fabrication. Technology Computer-Aided Design (TCAD) Sentaurus by Synopsys is one of the leading simulation platforms used worldwide in both academic and industrial semiconductor research for this purpose [151-153].

In the case of SOI VSTB FETs, TCAD Sentaurus helps in thoroughly investigating the impact of structural innovations, material choices, and real-device effects on device performance. These devices combine the benefits of vertical architecture with SOI technology, making their simulation more complex due to factors such as buried oxide isolation, vertical thin channels, and strong electrostatic coupling from the gate.

The simulation process for SOI VSTB FET begins with constructing a detailed 2D or 3D model of the device using Sentaurus Structure Editor (SDE). This can be accomplished using a graphical interface or by writing detailed Tool Command Language (TCL) scripts, depending on the complexity of the structure and the user's preference [151]. For SOI VSTB FET, the device definition includes the vertical silicon fin, source and drain regions, gate stack (typically comprising a high-k dielectric and metal gate), shallow trench isolation (STI) regions, and the buried oxide (BOX) layer that separates the active device from the substrate.

Once the geometry is finalized and meshed properly for numerical stability, the simulation proceeds in Sentaurus Device (SDevice), where all relevant physical models are applied. Since SOI VSTB FET operates in the nanoscale regime, it requires the incorporation of advanced models such as mobility degradation, bandgap narrowing, quantum corrections, generation-recombination mechanisms, impact ionization, and self-heating effects (SHE). These models ensure that the simulation captures realistic behaviors such as short-channel effects (SCEs), carrier mobility variations, thermal limitations, and parasitic effects that typically occur in SOI-based vertical devices [151-153].

One of the major advantages of using TCAD for device studies is the ability to simulate electrostatic potential distribution, carrier concentration profiles, electric field strengths, and heat dissipation patterns across the complex vertical architecture. Additionally, TCAD allows parametric sweeps of key design variables such as fin width, gate length, oxide thickness, and BOX layer thickness, enabling systematic optimization of the structure for desired electrical characteristics like higher ON-state current (I_{ON}), lower OFF-state leakage (I_{OFF}), improved sub-threshold swing (SS), and reduced DIBL [152].

For analyzing the thermal reliability, self-heating models are included to study how lattice temperature increases during device operation influence electrical characteristics like threshold voltage shift, degradation of carrier mobility, and thermal runaway scenarios [151]. Moreover, Sentaurus Visual and Inspect modules help visualize these parameters spatially for better insight and interpretation of simulation results.

TCAD Sentaurus also enables process simulations using Sentaurus Process, making it possible to simulate realistic fabrication flows such as shallow trench isolation (STI) formation, gate stack deposition, doping profiles for source/drain regions, and annealing treatments [153]. This integration of device and process simulation creates a holistic framework for designing advanced structures like SOI VSTB FETs with better predictability and reliability.

As such, TCAD Sentaurus provides a powerful and flexible environment for designing, modeling, and optimizing devices. It helps researchers address critical challenges such as short-channel effects, self-heating, interfacial traps, and material engineering by offering physics-based solutions through simulation. For next-generation CMOS applications, this simulation-driven approach greatly reduces development costs, saves time, and provides critical insights that would be difficult or expensive to obtain through physical prototyping alone.

2.6 Summary

The continuous progress of the semiconductor industry has been largely driven by the success of planar MOSFETs, guided by Moore's law and Dennard's scaling principle. However, as device dimensions shrank to the nanoscale, these conventional structures faced severe limitations due to short-channel effects (SCEs), leading to the need for new device architectures. Among various 3D structures, the SOI VSTB FET has emerged as one of the promising candidates. By combining the benefits of SOI technology with the vertical thin-body architecture, SOI VSTB FETs offer superior electrostatic control, reduced leakage current, and improved mechanical stability. This chapter presented a comprehensive survey of the evolution of MOS devices, highlighting key technological milestones that led to the development of SOI VSTB FETs. The fabrication process of these devices integrates advanced techniques, utilizing the buried oxide layer for isolation and shallow trench isolation (STI) walls to provide mechanical support for ultra-thin fins. Analytical models of key electrical parameters such as V_{TH} and surface potential were also discussed, providing deeper insight into the device physics. To fully evaluate the potential of SOI VSTB FETs, various performance aspects were explored, including electrical behavior, RF/analog characteristics, linearity, thermal reliability, and radiation resilience. Real-device effects like interfacial traps, architectural stress, self-heating, and noise were considered to reflect practical conditions. Material engineering approaches, including the use of silicon-germanium ($Si_{1-x}Ge_x$) channels and high-k gate dielectrics, were shown to offer additional performance benefits. The use of TCAD Sentaurus as a simulation platform enabled detailed study of the interplay between structural parameters and operational outcomes. Overall, SOI VSTB FETs present a compelling solution for next-generation electronic applications, offering advantages in terms of scalability, efficiency, and reliability. However, further research is essential to address remaining challenges in thermal management, process integration, and radiation hardening, ensuring their suitability for both advanced computing and mission-critical environments like space and defense systems.

References

[1] T. Lin et al., "J. S. Kilby, "Miniaturized electronic circuits [US Patent. No. 3,138, 743), *in IEEE Solid-State Circuits Society Newsletter*, vol. 12, no. 2, pp. 44-54, Spring 2007, doi: 10.1109/N-SSC.2007.4785580.

[2] C. Y. Wong, J. Y. Sun, Y. Taur, C. S. Oh, R. Angelucci and B. Davari, "Doping of 1n/sup / and p/sup +/ polysilicon in a dual-gate CMOS process, *Tech-nical Digest., International Electron Devices Meeting*, 1988, pp. 238-241, doi: 10.1109/IEDM.1988.32800

[3] R. N. Noyce, "Semiconductor Device-and-Lead Structure, *Reprint of U.S. Patent* 2,981.877 (Issued April 25, 1961. Filed July 30, 1959), in IEEE Solid-State Cir-cuits Society Newsletter, vol. 12, no. 2, pp. 34-40, Spring 2007, doi: 10.1109/N-SSC.2007.4785577.

[4] C. P. Wong, 'Recent advances in hermetic equivalent flip-chip hybrid IC pack-aging of microelectronics", *Materials Chemistry and Physics*, vol. 42, pp. 25-30, 1995

[5] D. Kahng and. M. Atalla, "Silicon-silicon dioxide field induced Surface Devices", *IRE-AIEE Solid-State Device Research Conference*, pp. 13 15, 1960

[6] D. Kahng, 'A historical perspective on the development of MOS transistors and related devices," *in IEEE Transactions on Electron Devices*, vol. 23, no. 7, pp. 655-657, July 1976, doi: 10.1109/T-ED.1976.18468

[7] M. Atalla, "Semiconductor devices having dielectric coatings." *U. S. Patent 3206670A*, Sept. 14, 1965. https://patents.google.com/patent/US3206670

[8] K. Dawon, "Electric field controlled semiconductor device." *U.S. Patent 3102230A*, Aug. 27, 1963. https://patents.google.com/patent/US3102230

[9] S.L. Moskowitz, "Advanced materials innovation: managing global technology in the 21st century," *John Wiley Sons*, pp. 165-167, 2016.

[10] J. Butrica, "NASA's role in the manufacture of integrated circuits" in His torical Studies in the Societal Impact of Spaceflight, *NASA*, PP149-250, ISBN 978-1-62683-027-1, 2015.

[11] D. L. Morton and J. Gabriel, "Electronics: The Life Story of a Technology, *Baltimore, MD, USA:Johns Hopkins Univ. Press*, 2007.

[12] L. T. Harrison, "Current Sources and Voltage References: A Design Reference for Electronics Engineers," Elsevier. p. 185. ISBN 978-0-08-045555-6, 2005.

[13] G. E. Moore, 'Cramming more components onto integrated circuits", *Proc. IEEE*, vol. 86, no. 1, pp. 82-85, Jan. 1998.

[14] G.-Q. Zhang, "More than Moore' The changing international landscape strat-egy and solutions of micro/nanoelectronics", *Proc. 8th Int. Conf. Electron. Packag. Technol.*, pp. 1, 2007.

[15] G. Q. Zhang, F. van Roosmalen and M. Graef, 'The paradigm of 'more than Moore', *Proc. 6th Int. Conf. Electron. Packag. Technol.*, pp. 17-24, 2005.

[16] G. Q. Zhang, M. Graef and F. van Roosmalen, 'The rationale and paradigm of 'more than Moore', *Proc. 56th Electron. Compon. Technol. Conf.*, pp. 151-157, 2006.

[17] M. M. Waldrop, 'More than Moore', *Nature*, vol. 530, pp. 144-147, Feb. 2016.

[18] R. R. Schaller, 'Moore's law: Past present and future', *IEEE Spectr.*, vol. 34, no. 6, pp. 52-59, Jun. 1997. doi: 10.1109/6.591665.

[19] T J. Misa, 'Understanding 'How Computing Has Changed the World', *in IEEE Annals of the History of Computing*, vol. 29, no. 4, pp. 52-63, Oct-Dec. 2007. doi: 10.1109/MAHC.2007.4407445.

[20] E Mollick, "Establishing Moore's Law." *in IEEE Annals of the History of Com-puting*, vol. 28, no. 3, pp. 62-75, July-Sept. 2006, doi: 10.1109/MAHC.2006.45.

[21] J, Koomey. S. Berard, M. Sanchez and H. Wong, 'Implications of His-torical Trends in the Electrical Efficiency of Computing, *in IEEE Annals of the History of Computing*, vol. 33, no. 3, pp. 46-54, March 2011, doi: 10.1109/MAHC.2010.28.

[22] D. Brock and G. Moore, Understanding Moore's Law: Four Decades of Inno-vation. *Chemical Heritage Foundation*, 2006.

[23] M. Lundstrom, 'Moore's law forever?", *Science*, vol. 299, no. 5604, pp. 210-211. 2003.

[24] B. G. Streetman and S. Banerjee, Solid State Electronic Devices, NJ, Upper Saddle River: Prentice-Hall, 2000.

[25] M. Bohr, 'A 30 Year Retrospective on Dennard's MOSFET Scaling Paper,' *in IEEE Solid-State Circuits Society Newsletter*, vol. 12, no. 1, pp. 11-13, Winter 2007, doi: 10.1109/N-SSC.2007.4785534.

[26] E. P. DeBenedictis, 'It's Time to Redefine Moore's Law Again, *in Computer*, vol. 50, no. 2, pp. 72-75, Feb. 2017, doi: 10.1109/MC.2017.34.

[27] B. Shen, K. Wilcox, A. M. Keshavarzi and D. Antoniadis, 'EP1: Moore's law challenges below 10nm: Technology, design and economic implications,' *IEEE International Solid-State Circuits Conference (ISSCC) Digest of Technical Papers*, 2015, pp. 1-1, doi: 10.1109/ISSCC.2015.7063150.

[28] M. B. Taylor, 'A Landscape of the New Dark Silicon Design Regime,' in IEEE Micro, vol. 33, no. 5, pp. 8-19, Sept.-Oct. 2013, doi: 10.1109/MM.2013.90.

[29] L. Kang, "The Art of Scaling: Distributed and Connected to Sustain the Golden Age of Computation, *IEEE International Solid-State Circuits Conference (ISSCC)*, 2022, pp. 25-31, doi: 10.1109/ISSCC42614.2022.9731536.

[30] D. M. Fleetwood. "Radiation Effects in a Post-Moore World." *in IEEE Trans actions on Nuclear Science*, vol. 68, no. 5, pp. 509-545, May 2021, doi: 10.1109/TNS.2021.3053424.

[31] B. Hendrickson, 'The Day After Tomorrow: The Looming Post-Exascale Cri-sis, *2018 IEEE International Parallel and Distributed Processing Symposium (IPDPS)*, 2018, pp. 710-710, doi: 10.1109/IPDPS.2018.00080

[32] A. Mehrabian, V. J. Sorger and T. El-Ghazawi, 'A Design Methodology for Post-Moore's Law Accelerators: The Case of a Photonic Neuromor-phic Processor, *2020 IEEE 31st International Conference on Application-specific Systems, Architectures and Processors (ASAP),* 2020, pp. 113-116, doi: 10.1109/ASAP49362.2020.00028.

[33] P. A. Gargini, 'Roadmap evolution: From NTRS to ITRS from ITRS 2.0 to IRDS", *Proc. Int. Conf. Extreme Ultraviolet Lithography*, vol. 10450, pp. 191-205, 2017.

[34] F. Clermidy et al., "Technology scaling: the CoolCube paradigm', SOI-3D-Subthreshold Microelectron. *Technol. Unified Conf. S3S IEEE*, pp. 1-4, 2015.

[35] Panagiotis Chaourani, Dimitrios Stathis, Saul Rodriguez, Per-Erik Hellström, Ana Rusu, 'A Study on Monolithic 3-D RF/AMS ICs: Placing Digital Blocks Under Inductors, *2018 IEEE SOI-3D-Subthreshold Microelectronics Technol-ogy Unified Conference (S3S)*, pp.1-3, 2018.

[36] M. Brocard, G. Berhault, S. Thuries, F. Clermidy, P. Batude, C. Fenouillet-Beranger, L. Brunet, F. Andrieu, F. Deprat, J. Lacord, O. Rozeau, G. Cibrario, O. Billoint, "Impact of intermediate BEOL technology mances of 3D VLSI, *2016 46th European Solid-State Device Research Confer-ence (ESSDERC)*, pp.218-221, 2016.

[37] M. B. Healy and S. K. Lim, "A study of stacking limit and scaling in 3D ICs: an interconnect perspective, *2009 59th Electronic Components and Technology Conference*, 2009, pp. 1213-1220, doi: 10.1109/ECTC.2009.5074166.

[38] Mario Donato Marino, 'Architectural Impacts of RFiop: RF to Address I/O Pad and Memory Controller Scalability", *IEEE Transactions on Very Large Scale Integration (VLSI) Systems*, vol.26, no.8, pp.1494-1507, 2018.

[39] Srivatsa Srinivasa, Xueqing Li, Meng-Fan Chang, John Sampson, Sumeet Ku-mar Gupta, Vijaykrishnan Narayanan, 'Compact 3-D-SRAM Memory With Concurrent Row and Column Data Access Capability Using Sequential Mono-lithic 3-D Integration', *IEEE Transactions on Very Large Scale Integration (VLSI) Systems,* vol.26, no.4, pp.671-683, 2018.

[40] Matthew Hogan, Dusan Petranovic, 'Robust verification of 3D-ICS: Pros, cons and recommendations', 2009 IEEE International Conference on 3D System In-tegration, pp. 1-6, 2009.

[41] J.-P. Colinge, 'Multiple gate silicon-on-insulator MOS transistors", Proc. Mi-croelectronics Technology and Dives - SBMICRO 2003, pp. 1-17, 2003-09 (2003)

[42] J.-P. Colinge, 'Recent advances in SOI technology, Proceedings of 1994 IEEE International Electron Devices Meeting, 1994, pp. 817-820, doi: 10.1109/IEDM.1994.383299.

[43] Jong-Tac Park and J.-P. Colinge, 'Multiple-gate SOI MOSFETS: device design guidelines, *in IEEE Transactions on Electron Devices*, vol. 49, no. 12, pp. 2222-2229. Dec. 2002, doi: 10.1109/TED.2002.805634.

[44] J.-P. Raskin, Tsung Ming Chung. V. Kilchytska, D. Lederer and D. Flandre, Analog/RF performance of multiple gate SOI devices: wideband simulations and characterization,' *in IEEE Transactions on Electron Devices*, vol. 53, no. 5, pp. 1088-1095, May 2006, doi: 10.1109/TED.2006.871876.

[45] S. De Franceschi et al., 'SOI technology for quantum information processing, *2016 IEEE International Electron Devices Meeting (IEDM)*, 2016, pp. 13.4.3 13.4.4, doi: 10.1109/IEDM.2016.7838409.

[46] H. Chi, Y. Wang, L. A. De Barros Naviner and W. Zhao, 'Robust Ultra-Lon Power Non-Volatile Logic-in-Memory Circuits in FD-SOI Technology,' *in IEE Transactions on Circuits and Systems I: Regular Papers*, vol. 64, no. 4, пр 847-857, April 2017, doi: 10.1109/TCSI.2016.2621344.

[47] X. J. Zhou, O. D. Patterson, W. h. Lee, H. H. Kang and R. Hah 'Characterization of contact module failure mechanisms for SOI technol ogy using E-beam inspection and in-line TEM. *2010 IEEE/SEMI Advancel Semiconductor Manufacturing Conference (ASMC)*, 2010, pp. 270-276, dol 10.1109/ASMC.2010.5551467.

[48] P. Dainesi et al., 'CMOS compatible fully integrated Mach-Zehnder interfer ometer in SOI technology,' *in IEEE Photonics Technology Letters*, vol. 12, n 6, pp. 660-662, June 2000, doi: 10.1109/68.849076.

[49] M. Jaffe et al., 'Improvements in SOI technology for RF switches, *2015 IEEE 15th Topical Meeting on Silicon Monolithic Integrated Circuits in RF Systems, 2015, pp. 30-32, doi: 10.1109/SIRF.2015.7119865.*

[50] D. Lederer and J. Raskin, 'RF Performance of a Commercial SOI Technol ogy Transferred Onto a Passivated HR Silicon Substrate, *in IEEE Trans actions on Electron Devices*, vol. 55, no. 7, pp. 1664-1671, July 2008, doi: 10.1109/TED.2008.923564

[51] V. Koldiaev, R. Pirogova, "Vertical super-thin body semiconductor on diele tric wall devices and methods of their fabrication," *U.S. Patent* 8,796,085 B2,5 Aug 2014.

[52] R. Pirogova, J. Wolf and V. Koldiaev. 'Investigation of novel vertical super-thin body silicon MOSFET performance advantages. *2016 IEEE Silicon Nanoelec tronics Workshop (SNW)*, 2016. pp. 1-2, doi: 10.1109/SNW 2016.7578012.

[53] N. P. Maity, R. Maity. S. Maity, S. Baishya, "Comparative analysis of the quan-tom FinFET and trigate FinFET based on modelling and simulation," *J. Com-pul.Electron.*, vol. 18, pp. 492-499, 2019. https://doi.org/10.1007/s10825-018-01294-z

[54] P Salua, P. Banerjee, S.K. Sarkar, "2D modeling based comprehensive anal-ysis of short channel effects in DMG strained VSTB FET." *Superlattices Mi-crestruct.*, vol. 118, pp. 16-28, 2018.

[55] S. Roy, A. Chatterjee, D. K. Sinha, R. Pirogova and S. Baishya, "2-D Analytical Modeling of Surface Potential and Threshold Voltage for Vertical Super-Thin Body FET,' *in IEEE Transactions on Electron Devices,* vol. 64, no. 5, pp. 2106-2112, May 2017, doi: 10.1109/TED.2017.2687465.

[56] R. R. Thakur and P, Singh, "Q-FinFET: the next generation FinFET," *J. Nanoelectron. Optoelectron.*, 2019, vol. 14, pp. 92-8.

[57] T. E. Rudenko, A. N. Nazarov and V. S. Lysenko, "The advancement of silicon-on-insulator (SOI) devices and their basic properties, *Semiconductor Physics Quantum Electronics Optoelectronics*, vol. 23, no. 3, 2020.

[58] X. Zhang, D. Connelly, H. Takeuchi, M. Hytha, R. J. Mears and T. K. Liu, "Comparison of SOI Versus Bulk FinFET Technologies for 6T-SRAM Voltage Scaling at the 7-/8-nm Node,' *in IEEE Transactions on Electron Devices*, vol. 64, no. 1, pp. 329-332, Jan. 2017. doi: 10.1109/TED.2016.2626397.

[59] Y. Sun, X. Yu, R. Zhang, B. Chen, and R. Cheng, "The past and future of multi-gate field-effect transistors: Process challenges and reliability issues" *Journal of Semiconductors*, 42(2), p.023102.

[60] C. Riddet, C. Alexander, A. R. Brown, S. Roy and A. Asenov, 'Simulation of "Ab Initio" Quantum Confinement Scattering in UTB MOSFETs Using Three-Dimensional Ensemble Monte Carlo,' *in IEEE Transactions on Electron Devices*, vol. 58, no. 3, pp. 600-608, March 2011, doi: 10.1109/TED.2010.2095422.

[61] N. P. Maity, R. Maity. S. Maity, S. Baishya, "Comparative analysis of the quan-tom FinFET and trigate FinFET based on modelling and simulation," *J. Com-pul. Electron*, vol. 18, pp. 492-499, 2019. https://doi.org/10.1007/s10825-018-01294-z

[62] P Salua, P. Banerjee, S.K. Sarkar, "2D modeling based comprehensive anal-ysis of short channel effects in DMG strained VSTB FET." *Superlattices Mi-crestruct.*, vol. 118, pp. 16-28, 2018.

[63] S. Roy, A. Chatterjee, D. K. Sinha, R. Pirogova and S. Baishya, "2-D Analytical Modeling of Surface Potential and Threshold Voltage for Vertical Super-Thin Body FET,' *in IEEE Transactions on Electron Devices*, vol. 64, no. 5, pp. 2106-2112, May 2017, doi: 10.1109/TED.2017.2687465.

[64] R. R. Thakur and P, Singh, "Q-FinFET: the next generation FinFET," *J. Nanoelectron. Optoelectron.*2019, vol. 14, pp. 92-8.

[65] T. E. Rudenko, A. N. Nazarov and V. S. Lysenko, "The advancement of silicon-on-insulator (SOI) devices and their basic properties, *Semiconductor Physics Quantum Electronics Optoelectronics*, vol. 23, no. 3, 2020.

[66] X. Zhang, D. Connelly, H. Takeuchi, M. Hytha, R. J. Mears and T. K. Liu, "Comparison of SOI Versus Bulk FinFET Technologies for 6T-SRAM Voltage Scaling at the 7-/8-nm Node,' *in IEEE Transactions on Electron Devices*, vol. 64, no. 1, pp. 329-332, Jan. 2017. doi: 10.1109/TED.2016.2626397.

[67] Y. Sun, X. Yu, R. Zhang, B. Chen, and R. Cheng, "The past and future of multi-gate field-effect transistors: Process challenges and reliability issues" *Journal of Semiconductors*, 42(2), p.023102.

[68] C. Riddet, C. Alexander, A. R. Brown, S. Roy and A. Asenov, 'Simulation of "Ab Initio" Quantum Confinement Scattering in UTB MOSFETs Using Three-Dimensional Ensemble Monte Carlo,' *in IEEE Transactions on Electron Devices*, vol. 58, no. 3, pp. 600-608, March 2011, doi: 10.1109/TED.2010.2095422

[69] A. Malinowski, J. Chen, S. K. Mishra, S. Samavedam and D. Sohn, "What is Killing Moore's Law? Challenges in Advanced FinFET Technology Integration, *2019 MIXDES 26th International Conference 'Mixed Design of Integrated Circuits and Systems'*, 2019, pp. 46-51,

doi: 10.23919/MIXDES.2019.8787084.

[70] D. Hisamoto, T. Kaga, Y. Kawamoto and E. Takeda, "A fully depleted lean channel transistor (DELTA)-a novel vertical ultra thin SOI MOSFET. *Inter-national Technical Digest on Electron Devices Meeting.* 1989, PP. 833-836, dof 10.1109/IEDM.1989.74182.

[71] D. Hisamoto et al., 'A folded-channel MOSFET for deep-sub-tenth micron era, International Electron Devices Meeting 1998. Technical Digest (Cat. No.98CH36217), 1998, pp. 1032-1034, doi: 10.1109/IEDM.1998.746531.

[72] D. Hisamoto, 'A folded-channel MOSFET for deep-sub-tenth micron era', IEDM Tech. Dig., pp. 1032-1034, 1998.

[73] D. Hisamoto et al., 'FinFET-a self-aligned double-gate MOSFET scalable to 200 nm, *in IEEE Transactions on Electron Devices*, vol. 47, no. 12, pp. 2320-2325, Dec. 2000, doi: 10.1109/16.887014.

[74] Xuejue Huang et al., 'Sub 50-nm FinFET: PMOS, *International Electron Devices Meeting* 1999. Technical Digest (Cat. No.99CH36318), 1999, pp. 67-70, doi: 10.1109/IEDM.1999.823848.

[75] Xuejue Huang et al., 'Sub-50 nm P-channel FinFET,' *in IEEE Transactions on Electron Devices*, vol. 48, по. 5, pp. 880-886, May 2001, doi: 10.1109/16.918235

[76] M. Jurczak, N. Collaert, A. Veloso, T. Hoffmann and S. Biesemans, *'Review of FINFET technology*, 2009 IEEE International SOI Conference, 2009, pp. 1-4, doi: 10.1109/SOI.2009.5318794.

[77] R. S. Pal. S. Sharma and S. Dasgupta, "Recent trend of FinFET devices and its challenges: A review, *2017 Conference on Systems (ICEDSS)*, 2017, pp. 150-154. doi: 10.1109/ICEDSS.2017.8073675.

[78] Bu Liang Yang et al, 5nm-gate nanowire FinFET, *Digest of Technical Papers 2001 Symposium on VLSI Technology*, 2001., 2004, pp. 196-197, doi: 10.1109/VLSIT: 2004.1345476.

[79] Jacopo lannacci, H. Vincent Poor, 'Review and Perspectives of Micro/Nano Texhnologies, *as Key-Fatablers of 6G, IEEE Access*, vol. 10, pp.55428-55458,

[80] Matsukawa et al, Comprehensive analysis of variability sources of FinFET Characteristics. *2009 Symposium on VLSI Technology*, 2009, pp. 118-119.

[81] International Technology Roadmap for Semiconductors 2.0, 2015. On-fine] Available: https://www.semiconductors.org/wp-content/uploads/2018/06/4-2015-ITRS-2.0_0SC-.pdf. (Accessed: 5ãĂŞDecember-2017].

[82] A Godoy, J. A. Lpez-Villanueva, J. A. Jimnez-Tejada, A. Palma and F. Gmiz, A simple subthreshold swing model for short channel MOSFETs, *Solid-State Electron.* vol. 45, no. 3, pp. 391-397, Mar. 2001.

[83] A. Tsormpatzoglou, D. H. Tassis, C. A. Dimitriadis, G. Ghibaudo, G. Pananakakis and N. Collaert, 'Analytical modeling for the current-voltage char-acteristics of undoped or lightly-doped symmetric double-gate MOSFETs', *Mi-croelectron. Eng.*, vol. 87, no. 9, pp. 1764-1768, Nov. 2010.

[84] B. Mukhopadhyay, A. Biswas, P. K. Basu, G. Eneman, P. Verheyen, E. Simoen, et al., "Modelling of threshold voltage and subthreshold slope of strained-Si MOSFETs including quantum effects', *Semicond. Sci. Technol.*, vol. 23, no. 9, pp. 095017, Sep. 2008.

[85] S. R. N. Yun. C. G. Yu, J. T. Park, C.-W. Lee, D. Lederer. A. Afzalian, et al., "A quantum definition of threshold voltage in MuGFETs", *Proc. IEEE Int. SOI Conf. Proc.*, pp. 137-138, 2007.

[86] W. Wang, H. Lu, J. Song, S.-H. Lo and Y. Taur, "Compact modelling of quantum effects in symmetric double-gate MOSFETs', *Microelectron. Eng.*, vol. 41, no. 10, pp. 688-692, Oct. 2010.

[87] Y.-C. Eng et al., Importance of VDIBLSS/(I_{on}/I_{oft}) in evaluating the per formance of n-channel bulk FinFET devices,' *in IEEE Journal of the Electron Devices Society*, vol. 6, pp. 207-213, 2018, doi: 10.1109/JEDS.2018.2789922.

[88] B. Andreev, E.L. Titlebaum, E.G. Friedman, "Sizing CMOS invert ers with miller effect and threshold voltage variations," *Journal of Cir-cuits, Systems and Computers*, vol. 15, no. https://doi.org/10.1142/S021812660600314З 3. pp. 437-454, 2006,

[89] V. Vijayvargiya and S. K. Vishvakarma, "Effect of Drain Doping Profile on Double-Gate Tunnel Field-Effect Transistor and its Influence on Device RF Per-formance." *in IEEE Transactions on Nanotechnology*, vol. 13, no. 5, pp. 974-981. Sept. 2014, doi: 10.1109/TNANO.2014.2336812.

[90] A. Kranti and G. A. Armstrong, 'Nonclassical Channel Design in MOSFETs for Improving OTA Gain-Bandwidth Trade-Off,' *in IEEE Transactions on Circuits and Systems I: Regular Papers*, vol. 57, no. 12, pp. 3048-3054, Dec. 2010, doi 10.1109/TCSI.2010.2071470.

[91] R. K. Sharma and M. Bucher, 'Device Design Engineering for Optimum Analog/RF Performance of Nanoscale DG MOSFETs, in IEEE Transac-tions on Nanotechnology, vol. 11, no. 5, pp. 992-998, Sept. 2012, doi: 10.1109/TNANO.2012.2204439.

[92] Y. Pratap, S. Haldar, R. S. Gupta and M. Gupta, 'Performance Evaluation and Reliability Issues of Junctionless CSG MOSFET for RFIC Design,' *in IEEE Transactions on Device and Materials Reliability*, vol. 14, no. 1, pp. 418-425, March 2014, doi: 10.1109/TDMR.2013.2296524.

[93] M. K. Anvarifard and A. A. Orouji. 'Evidence for Enhanced Reliability in a Novel Nanoscale Partially-Depleted SOI MOSFET." *in IEEE Transactions on Device and Materials Reliability*, vol. 15. no. 4, pp. 536-542. Dec. 2015. doi: 10.1109/TDMR.2014.2328583

[94] N. B. Bousari, M. K. Anvarifard and S. Haji-Nasiri, 'Improving the electrical characteristics of nanoscale triple-gate junctionless FinFET using gate oxide engineering. *AEU Int. J. Electron. Commun*, vol. 108, pp. 226-234, Aug. 2019.

[95] S. K. Mohapatra, K. P. Pradhan, D. Singh and P. K. Sahu, "The Role of Geom-etry Parameters and Fin Aspect Ratio of Sub-20nm SOI-FinFET: An Analysis Towards Analog and RF Circuit Design,' *in IEEE Transactions on Nanotechnology*, vol. 14, no. 3, pp. 546-554, May 2015, doi: 10.1109/TNANO.2015.2415555.

[96] T. Ohtou, T. Saraya and T. Hiramoto, 'Variable-Body-Factor SOI MOSFET With Ultrathin Buried Oxide for Adaptive Threshold Voltage and Leakage Con-trol. *in IEEE Transactions on Electron Devices*, vol. 55, no. 1, pp. 40-47, Jan. 2008, doi: 10.1109/TED.2007.912612.

[97] Suruchi Sharma, Rikmantra Basu, Baljit Kaur, "Insights into temperature influence on analog/RF and linearity performance of a Si/Ge heterojunction asymmetric double gatedopingless TFET," *Applied Physics A*, vol. 127, pp. 392, 2021. https://doi.org/10.1007/s00339-021-04541-6

[98] Y. Tsividis, C. McAndrew, "The MOS Transistor," *Oxford University Press*, pp. 69-70, 2013.

[99] C. H. Lee, T. Nishimura, K. Nagashio, K. Kita and A. Toriumi, "High-Electron-Mobility Ge/GeO2 n-MOSFETs With Two-Step Oxidation, *in IEEE Trans-actions on Electron Devices*, vol. 58, no. 5, pp. 1295-1301, May 2011, doi: 10.1109/TED.2011.2111373.

[100] N. Miyata, "Study of direct-contact HfO2 /Si interfaces, Materials, vol. 5, no. 3, pp. 512-527, Mar. 2017. https://doi.org/10.3390/ma5030512

[101] R. Goswami, B. Bhowmick, S. Baishya, "Electrical noise in circular gate tunnel FET in presence of interface traps," *Superlattices and Microstructures*, vol. 86, pp. 342-354, 2015.

[102] S.Y. Wu, "Theory of generation recombination noise in MOS transistors," *Solid State Electron.* vol.11, pp.25-32, 1968

[103] P. Anandan, A. Nithya, N. Mohankumar, "Simulation of flicker noise in gate all-around silicon nanowire MOSFETs including interface traps," *Microelectron ics Reliability*, vol. 54, pp. 2723-2727, 2014.

[104] L.K.J. Vandamme. F.N. Hooge, "What do we certainly know about 1/f noise in MOSTs?" *IEEE Transactions on Electron Devices*, vol. 55.issue 11, pp. 3070-3085 (2008). https://doi.org/10.1109/TED.2008.20051 67

[105] S.-C. Hsu and Y. Li, Electrical characteristic fluctuation of 16-nm-gate high k/metal gate bulk FinFET devices in the presence of random interface traps *Nanosc.Res. Lett.* vol. 9, no. 1, pp. 633, Dec. 2014.

[106] J. W. Lee, W. S. Yun and G. Ghibaudo, "Impact of trap localization on low. Frequency noise in nanoscale device', *J. Appl. Phys.*, vol. 115, no. 19, May 2014.

[107] J.-S. Yoon, 'DC performance variations of soi finfets with different silicide thickness", *Advances in Condensed Matter Physics*, vol. 2018, pp. 1-7, 2018. https://doi.org/10.1155/2018/2426863

[108] K. Sinha, "Three Dimensional Process Induced Stress in Nano-scale MOS FET using Embedded Source/Drain Technology," *International Conference on Advanced Materials and Energy Technology (ICAMET)*, 2014, December 17-19, IIEST Shibpur, Howrah, West Bengal, India.

[109] A. L. S. Loke. D. Yang, T. T. Wee, J. L. Holland, P. Isakanian, K. Rim, S. Yang, J. S. Schneider, G. Nallapati, S. Dundigal, H. Lakdawala, B. Amelifard. C. Lee, B. McGovern, P. S. Holdaway. X. Kong and B. M. Leary 'Analog/mixed-signal design challenges in 7-nm CMOS and beyond, *2018 IEEE Custom Integrated Circuits Conference (CICC)*, 2018, pp. 1-8, doi: 10.1109/CICC.2018.8357060.

[110] Kunal Sinha, Sanatan Chattopadhvav, Partha Sarathi Gupta and Hafizur Rahaman. "A technique to incorporate both tensile and compressive channel stress in Ge FinFET architecture", *Journal of Computational Electronics*, vol 16, no. 3, pp. 620-630, September 2017.

[111] K. Boncart, W. Riess and A. M. Imescu, 'Lateral Strain Profile as Key Tech-nology Booster for All-Silicon Tunnel FETs,' *in IEEE Electron Device Letters*, col. 30, no. 6. pp. 656-658, June 2009. doi: 10.1109/LED.2009.2018127.

[112] H Wang, X. Jiang, N. Xu, G. Han, Y. Hao, S.-S. Li and D. Esseni, 'Revised Analysis of Design Options and Minimum Subthreshold Swing in Piezoelectric FinFETs, in IEEE Electron Device Letters, vol. 39, no. 3, pp. 444-447, March 2018, doi: 10.1109/LED.2018.2791987.

[113] H Kim, G Roh, Y. K. Kwon and B. Kang, "Channel-length dependence of mechanical stress effect by hybrid shallow trench isolation on NBTI degradation of HfSION/SiO2 p-channel MOSFETs with strained Si/SiGe channel" *Micro-electron. Reliability*, 2019.

https://doi.org/10.1016/j.microrel.2019.113424

[114] Xiangchen Chen and Cher Ming Tan, 'Modeling and analysis of gate-all-around silicon nanowire FET, *Microelectronics Reliability*, vol. 54, pp. 1103-1108. July 2013. https://doi.org/10.1016/j.microrel.2013.12.009

[115] Q. Guo, Z. Di, M. G. Lagally and Y. Mei, 'Strain engineering and mechanical assembly of silicon/germanium nanomembranes", *Mater. Sci. Eng. R Rep.*, vol. 128, pp. 1-31, Jun. 2018. https://doi.org/10.1016/j.mser.2018.02.002

[116] P. Yang, "Effect of Gate-Line-End-Induced Stress and Its Impact on Device's Characteristics in FinFETs,' *in IEEE Electron Device Letters*, vol. 37, no. 7, pp. 910-912, July 2016, doi: 10.1109/LED.2016.2565901.

[117] G. He and Z. Sun, "High-k Gate Dielectrics for CMOS Technology," *New York, NY, USA:Wiley*, 2012.

[118] A. A. Demkov and A. Navrotsky, Materials Fundamentals of Gate Dielectrics *(Springer, Dordrecht, The Netherlands*, 2005) p. 403ff.

[119] H. R. Huff and D. C. Gilmer, "High Dielectric Constant Materials," *Berlin, Germany: Springer-Verlag*," 2005.

[120] I. De, D. Johri, A. Srivastava, C.M. Osburn, "Impact of gate work function on device performance at the 50 nm technology node," *Solid-State Electronics*, vol. 44, pp. 1077–1080, 2000.

[121] F. Crupi et al., "Impact of the interfacial layer on the low-frequency noise (1/f) behavior of MOSFETs with advanced gate stacks," *IEEE Electron Device Letters*, vol. 27, no. 8, pp. 688–691, Aug. 2006.

[122] W. S. Ho et al., 'Thermal oxide, Al_2O_3 and amorphous-Si passivation layers on silicon, *2010 35th IEEE Photovoltaic Specialists* 003166, doi: 10.1109/PVSC.2010.5615938. sts Conference, 2010, pp. 003163

[123] M.H. Cho, D.H. Ko, Y.G. Choi, K. Jeong. D.Y. Noh. H.J. Kim, C.N. Whang. "Structural and electrical characteristics of YoO, films grown on o idized Si(100) surface, *J. Vac. Sci. Technol. A*. vol. 19. pp. 192-199, 2001 https://doi.org/10.1116/1.1331296

[124] I. De, D. Johri, A. Srivastava, C.M. Osburn. "Impact of gate work function on device performance at the 50 nm technology node." *Solid State Electronics* vol. 44, PP. 1077-1080, 2000. https://doi.org/10.1016/S0038-1101(99)00323-8

[125] F. Crupi et al., 'Impact of the interfacial layer on the low-frequency noise (1/1) behavior of MOSFETs with advanced gate stacks," *in IEEE Electron Device Letters*, vol. 27, no. 8, pp. 688-691, Aug. 2006, doi: 10.1109/LED.2006.87902

[126] A. S. Oates, 'Reliability issues for high-k gate dielectrics, *IEEE In-ternational Electron Devices Meeting* 2003, 2003, pp. 38.2.1-38.2.4, doi: 10.1109/IEDM.2003.1269429.

[127] Y. Wu, L. Chen, R. Lyu, M. Li and H. Wu, "Tetragonal ZrO2/Al2O3 Stack as High-Gate Dielectric for Si-Based MOS Devices," *in IEEE Electron Device Let-ters*, vol. 31, no. 9, pp. 1014-1016, Sept. 2010, doi: 10.1109/LED.2010.2053191

[128] N. R. Mohapatra, M. P. Desai, S. G. Narendra and V. R. Rao, "The effect of high-K gate dielectrics on deep submicrometer CMOS device and circuit performance', *IEEE Trans. Electron Devices*, vol. 49, no. 5, pp. 826-831, May 2002. doi: 10.1109/16.998591.

[129] C. R. Manoj and V. R. Rao, 'Impact of High-k Gate Dielectrics on the De vice and Circuit Performance of Nanoscale FinFETs." *in IEEE Electron Device Letters*, vol. 28, no. 4, pp. 295-297. April 2007. doi: 10.1109/LED.2007.892365

[130] T. T. Mnatsakanov. I. L. Rostovtsev and N. I. Philatov, 'Investigation of the effect of nonlinear physical phenomena on charge carrier transport in semicon-ductor devices", *Solid State Electro.*, vol. 30, no. 6, pp. 579-585, 1987.

[131] M. Lamdstrom, "Fundamentals of Carrier Transport," *U.K., Cam-Bridge: Cambridge Univ. Press*, 2000.

[132] T. Fang, R. Wang, H. Xing, S. Rajan and D. Jena, 'Effect of Optical Phonon Scattering on the Performance of GaN Transistors,' *in IEEE Electron Device Letters*, vol. 33, no. 5, pp. 709-711, May 2012, doi: 10.1109/LED.2012.2187169.

[133] S. -S. Lu. T.-W. Chen, H. C. Chen and C. Meng, "The origin of the kink phenomenon of transistor scattering parameter S/sub 22/, *in IEEE Transac-tions on Microwave Theory and Techniques*, vol. 49, no. 2, pp. 333-340, Feb. 2001, doi: 10.1109/22.903094.

[134] G. K. Wachutka, "Rigorous thermodynamic treatment of heat generation and conduction in semiconductor device modeling, *in IEEE Transactions on Computer-Aided Design of Integrated Circuits and Systems*, vol. 9, no. 11, pp. 1141-1149, Nov. 1990, doi: 10.1109/43.62751.

[135] K. A. Jenkins and K. Rim, 'Measurement of the effect of self-heating in strained-silicon MOSFETs,' *in IEEE Electron Device Letters*, vol. 23, по. 6, pp. 360-362, June 2002, doi: 10.1109/LED.2002.1004235.

[136] T. Takahashi, S. Oda, and K. Uchida "Methodology for evaluating operation temperatures of fin-type field-effect transistors connected by interconnect wires," *Proc. Jpn. J. Appl. Phys.*, vol. 52, no. 6R, May 2013, Art. no. 064203.

[137] Myeong, D. Son, H. Kim, M. Kang, J. Jeon and H. Shin, "Thermal-Aware Shallow Trench Isolation Design Optimization for Minimizing Ioff in Various Sub-10-nm 3-D Transistors,' *in IEEE Transactions on Electron Devices*, vol. 66, no. 1, pp. 647-654, Jan. 2019, doi: 10.1109/TED.2018.2882577

[138] T.Takahashi, T. Matsuki, T. Shinada, Y. Inoue, and K. Uchida "Comparison self-heating effect (SHE) in short-channel bulk and ultra-thin BOX SOL MOS FETS: Impacts of doped well ambient temperature and SOI/BOX thicknesses on SHE," *in Proc. IEEE Int. Electron Devices Meeting, Washington, DC, USA*, 2013, pp. 7.4.1-7.4.4. doi: 10.1109/IEDM.2013.6724581.

[139] G. Kaushal, S. S. Rathod, S. Maheshwaram, S. K. Manhas, A. K. Saxena and S. Dasgupta, "Radiation Effects in Si-NW GAA FET and CMOS Inverter: A TCAD Simulation Study," *in IEEE Transactions on Electron Devices*, vol. 59, no. 5, pp. 1563-1566, May 2012, doi: 10.1109/TED.2012.2187656.

[140] P. S. Rajakumar and S. Satheesh Kumar, "A Comprehensive Review of Single Event Transients on Various MOS Devices," *in IEEE Access*, vol. 12, pp. 154760-154777, 2024, doi: 10.1109/ACCESS.2024.3483223.

[141] K. Aishwarya and B. Lakshmi, "Radiation study of TFET and JLFET-based devices and circuits: A comprehensive review on the device structure and sensitivity," *Radiat. Eff. Defects Solids*, vol. 178, nos. 3–4, pp. 229–257, 2022, doi: 10.1080/10420150.2022.2133708

[142] S. Tiwari, R. Saha, G. Gopal and T. Varma, "Sensitivity Analysis of Ge Source Double Gate TFET Under Heavy Ion Irradiation," *in IEEE Sensors Journal*, vol. 24, no. 22, pp. 36730-36736, 15 Nov.15, 2024, doi: 10.1109/JSEN.2024.3463870.

[143] S. E. Liu et al., 'Self-heating effect in FinFETs and its impact on devices reliability characterization.' *2014 IEEE International Reliability Physics Sym-posium*, 2014, pp. 4A.4.1-4A.4.4, doi: 10.1109/IRPS.2014.6860642.

[144] S. Lee et al., Experimental analysis and modeling of self heating effect in dielectric isolated planar and fin devices, *2013 Symposium on VLSI Technology*, 2013, pp. T248-T249.

[145] C. W. Chang, S. E. Liu, B. L. Lin, C. C. Chiu, Y.-H. Lee and K. Wu, "Thermal behavior of self-heating effect in FinFET devices acting on back-end interconnects, *2015 IEEE International Reliability Physics Symposium*, 2015, pp. 2F.6.1-2F.6.5, doi: 10.1109/IRPS.2015.7112696.

[146] J. Roig, D. Flores, M. Vellvehi, J. Rebollo and J. Millan, 'Reduction of self-heating on SOIM devices", *Microelectron. Reliab.* vol. 42, no. 1, pp. 61-66, Jan. 2002.

[147] L. Cai, W. Chen, G. Du, X. Zhang and X. Liu, 'Layout Design Correlated With Self-Heating Effect in Stacked Nanosheet Transistors, in IEEE Trans actions on Electron Devices, vol. 65, no. 6, pp. 2647-2653, June 2018, doi: 10.1109/TED.2018.2825498.

[148] M. A. Alarm, B. K. Mahajan, Y.-P. Chen, W. Ahn, H. Jiang and S. H. Shin, 'A Device-to-System Perspective Regarding Self-Heating Enhanced Hot Carrier Degradation in Modern Field-Effect Transistors: A Topical Review. *in IEEE Transactions on Electron Devices*, vol. 66, no. 11. pp. 4556-4565, Nov. 2019, dov 10.1109/TED.2019.2941445.

[149] O, Semenov, A. Vassighi and M. Sachdev, 'Impact of self-heating effect on long-term reliability and performance degradation in CMOS circuits, *in IEEE Transactions on Device and Materials Reliability*, vol. 6, no. 1, pp. 17-27, March 2006. doi: 10.1109/TDMR.2006.870340.

[150] AM. A. Belaid, K. Ketata, H. Maanane, M. Gares, K. Mourgues and J. Mar-con. "Analysis and Simulation of Self-Heating Effects on RF LDMOS Devices, *2005 International Conference On Simulation of Semiconductor Processes and Devices*, 2005, pp. 231-234, doi: 10.1109/SISPAD.2005.201515.

[151] "Sentaurus™ Device User Guide, Version M-2017.09", *Mountain View, CA, USA*, Sep. 2017.

[152] Y.-C. Wu and Y.-R. Jhan, 'Introduction of synopsys sentau-rus TCAD simulation" *in 3D TCAD Simulation for CMOS Na-noeletronic Devices, Singapore: Springer*, pp. 1-17, 2018, [online]

[153] Sentaurus™ Process Explorer User Guide Version R-2020.09', Sep. 2017

CHAPTER 3

Design, Optimization and Electrical Characterization of SOI VSTB FETs: A Comparative Study

3.1 Introduction

With the continuous downscaling of transistor dimensions to meet the increasing demand for faster and more energy-efficient electronic devices, conventional bulk MOSFETs are facing severe limitations due to short-channel effects (SCEs) [1-2]. Although FinFET and other multi-gate structures have demonstrated improved electrostatic integrity, they suffer from issues such as increased aspect ratio requirements, fabrication complexity, and mechanical instability of ultra-thin fins at nanometer scales [3–6]. To overcome these challenges, the Vertical Super-Thin Body (VSTB) FET was proposed, offering superior structural stability and better gate control through its vertical channel supported by shallow trench isolation (STI) dielectric walls [7-8]. However, conventional bulk VSTB FETs are limited by the presence of a direct body-to-substrate connection, leading to unwanted leakage currents and parasitic effects, particularly under aggressive scaling [7-9].

To mitigate these limitations, this chapter investigates a Silicon-On-Insulator SOI VSTB FET, where a buried oxide (BOX) layer is incorporated beneath the active silicon region to electrically isolate the device from the substrate [9]. This architecture significantly reduces substrate coupling, suppresses leakage, and enhances short-channel performance. Additionally, to further strengthen gate control and minimize leakage currents in scaled geometries, a high-k gate dielectric (HfO_2) is employed [9]. The use of HfO_2, owing to its higher dielectric constant, allows for improved gate capacitance without the need for a physically thinner oxide, thereby reducing gate tunneling leakage while maintaining strong electrostatic control [10].

This chapter presents a comprehensive comparative study between the SOI VSTB FET and the conventional bulk VSTB FET in terms of key electrical parameters, including threshold voltage (V_{th}), on-current (I_{on}), off-current (I_{off}), I_{on}/I_{off} ratio, subthreshold swing (SS), and drain-induced barrier lowering (DIBL) [14-15]. The aim is to establish the effectiveness of combining SOI architecture with high-k dielectric engineering to develop high-performance (HP) and low-power (LP) semiconductor devices suitable for advanced CMOS scaling nodes. The chapter organization is as follows: Section 3.2 describes the SOI VSTB FET structure along with the simulation setup, material specifications, and modeling considerations adopted for this study. Section 3.3 presents and discusses the detailed comparative results of SOI VSTB FET and bulk VSTB FET, focusing on electrical parameters such as threshold voltage, leakage current, on-state current, subthreshold swing (SS), and DIBL. Finally, Section 3.4 provides the summary and conclusion of the analysis carried out in this chapter.

3.2 Device Structure and TCAD Calibration

The three dimensional schematic diagrams of Bulk VSTB FET and SOI VSTB FET are shown in Fig. 3(a) and Fig. 3(b), respectively. Fig. 3(c) and Fig. 3(d) show the 2-D schematic top view and colour code, respectively, for the SOI-VSTB FET. The fabrication steps of such structures are same as discussed in [5]. The source, channel, drain, and substrate doping concentrations are 10^{19}, 10^{15}, 10^{17}, and 10^{15} cm^{-3}, respectively. The drain is weakly doped in comparison to the source to minimize the leakage current (I_{OFF}). A high k (ε_r = 22) dielectric oxide (HfO_2) with a thickness t_{ox}=2 nm with a TiN gate was added to get better results. We focussed on the performance of the device as specified by the ITRS standards in terms of per micrometre [6]. Descriptions of critical device parameters with their corresponding dimensions are listed in Table 3.1.

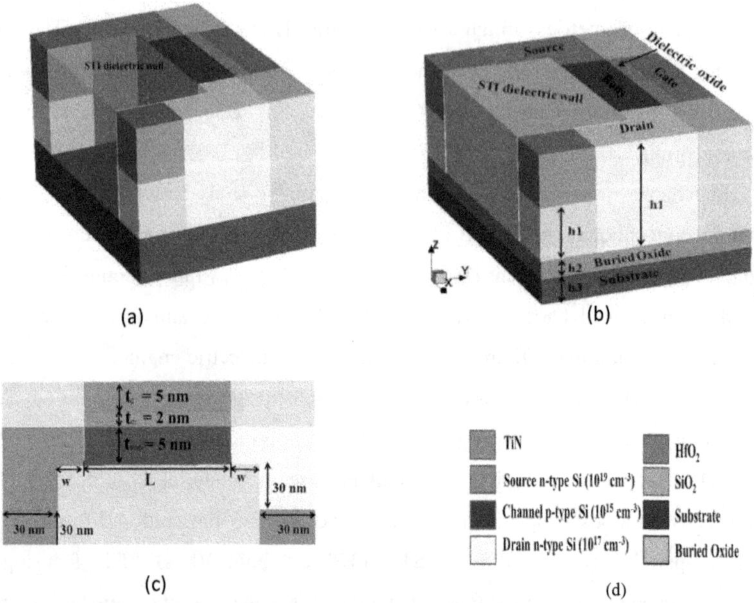

FIGURE 3.1 (a) 3D view of VSTB FET, (b) 3D view of SOI VSTB FET,(c). 2D top view, and (d) Regions name.

The device characteristics were simulated using three-dimensional Sentaurus TCAD simulation tool [11-12]. The drift-diffusion model was applied as a transport model [13-14]. Various physics-based models were triggered to depict realistic phenomena. Fermi-Dirac statistic model was activated so as to the effects of high doping concentration employed in the source and drain region are taken care of. The doping-dependent Masetti model, high-field saturation model, and the enhanced Lombardi model with high-k degradation were also activated for the dependency of carrier mobility on the device doping profile, high electric field, and high-*k* dielectric oxide, respectively [14]. The impact of strong electric field on carrier velocity was taken into consideration by triggering the velocity saturation model [14]. The physical dimensions of the Shallow Trench Isolation (STI) significantly impact device performance; specifically, a broader STI region can introduce mechanical stress and strain within the ultra-thin vertical channel [14]. To accurately capture these phenomena, this study utilized models for stress-induced electron mobility and stress-dependent velocity saturation [14], allowing for a comprehensive analysis of how stress alters device behavior [14-15]. Furthermore, the effects of strain were incorporated into the simulation by activating the deformation potential (DF) model.

Table 3.1 Description of each device parameters with their respective dimensions

Parameters	Dimensions (nm)
Body thickness, t_{body}	5
Gate thickness, t_g	5
Dielectric oxide thickness, t_{ox}	2
Channel length, L	21
Source and drain extension, w	4
SiO_2 layer height employed between source/drain/gate contact and substrate, h	15
Body height, h_1	35
Buried oxide, h_2	20
Substrate height, h_3	20

Within the deformation potential framework, specific flags were enabled to account for shifts in the conduction and valence band edges caused by strain [12, 14]. A correctly calibrated TCAD Sentaurus tool was used to carry out all the analyses. The experimental results [16] have been used in calibrating our TCAD tools. As shown in Fig. 3.2, our calibrated tools match well with the experimental data.

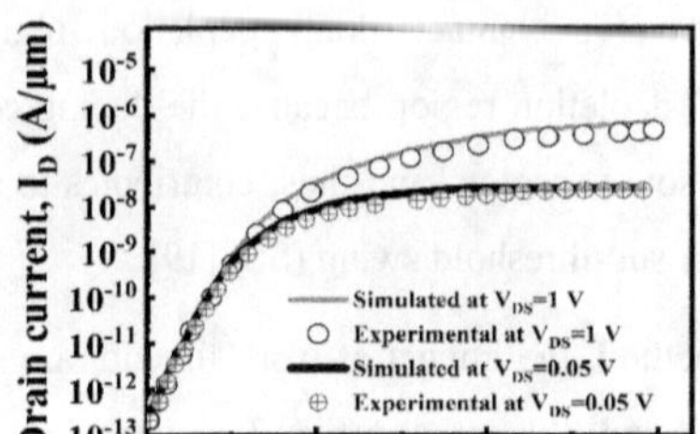

FIGURE 3.2 Calibration of the TCAD model against experimental data [16].

3.3 Results and Discussion

3.3.1 Optimization of device design parameters

The impacts of different design parameters like doping concentration, workfunction, channel length, oxide thickness, and aspect ratio (H/W) on the electrical characteristics of both the VSTB FET and SOI VSTB FET are discussed. Fig. 3.3(a) and Fig. 3.3(b) show the transfer characteristics of VSTB FET and SOI VSTB FET, respectively, for five doping profiles (DP1-DP5) listed in Table 3.2. The doping profile is optimized for a minimum leakage current. Gradual improvement in sub-threshold swing, Off and On state currents for both the bulk and SOI VSTB FET is evident when doping values in the drain region become lower from the Doping profile 1 (DP1) to Doping profile 4 (DP4). The doping concentration value of the source is again decreased in Doping profile 5 (DP5) to attain a low off-current without any considerable changes in on-current for both the structures. Furthermore, the corresponding increment in drain impedance for a lightly doped (LDD) drain degrades the on-current (I_{ON}) [17-19].

Table 3.2: Doping profile used in the simulation

Doping Profile (DP)	Source	Channel	Drain
DP1(cm^{-3})	10^{20}	10^{15}	10^{20}
DP2(cm^{-3})	10^{20}	10^{15}	10^{19}
DP3(cm^{-3})	10^{20}	10^{15}	10^{18}
DP4(cm^{-3})	10^{20}	10^{15}	10^{17}
DP5(cm^{-3})	10^{19}	10^{15}	10^{17}

SOI VSTB FET device exhibits better electrical characteristics. Most of the electric drain fields end up in the broader channel drain depletion region rather than in the channel/source-to-channel depletion region because the doping concentration of the drain region is lower than the source region and thus, contributes to a drastic reduction in the off-current(I_{OFF}) and better sub-threshold swing (SS) [19].

By using our proposed method, the impact of work function on the VSTB FET and SOI-VSTB FET is investigated and illustrated in Fig. 3.3(c) and Fig. 3.3(d), respectively. It is evident that the SOI VSTB FET is more vulnerable to work function (larger V_{th} variation), and this difference increases with increasing work function. This is due to the fact that as the metal gate work function increases, threshold voltage increases means at a high gate voltage channel is formed so off-state current reduces in bulk VSTB FET as well as in SOI VSTB FET. Still, SOI VSTB FET is more vulnerable because of additional suppression of off-state current due to the removal of source-substrate and drain-substrate p-n junction due to the presence of the buried oxide layer [20-22].

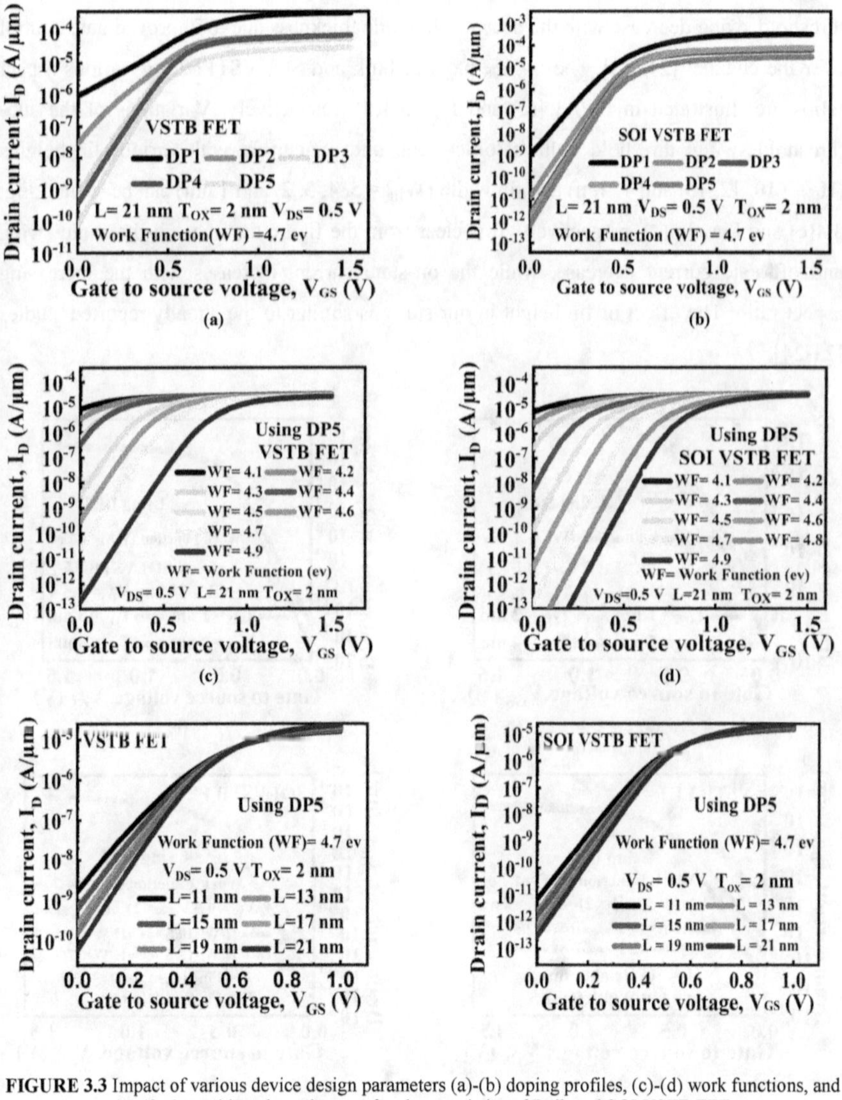

FIGURE 3.3 Impact of various device design parameters (a)-(b) doping profiles, (c)-(d) work functions, and (e)-(f) channel length on the transfer characteristics of Bulk and SOI VSTB FET

The impact of channel length (L_{ch}) on the transfer characteristics of bulk and SOI VSTB FET are shown in Fig. 3.3(e) and Fig. 3.3(f), respectively. From the plots, it is clear that the ON-current is hardly increased for both the devices; whereas, I_{OFF} significantly reduced especially in SOI VSTBFET as the L_{ch} is increased from 11 to 21 nm, due to the insertion of the buried oxide layer [21-22]. Fig. 3.4(a) and Fig. 3.4(b) show the I_D-V_{GS}

characteristics for varying oxide thickness (t_{ox}). It is observed that the off-current and sub-threshold swing decrease with the decrease in oxide thickness due to improved gate control over the channel [21, 23]. The I_D-V_{GS} plots of bulk and SOI VSTFET for various aspect ratios are illustrated in Fig. 3.4(c) and Fig. 3.4(d), respectively. Variations of the sub-threshold swing, threshold voltage, on-current, and off-current with various fin heights (H_{fin} = 10, 12, 15, and 17 nm) and fin width (W_{fin} = 5, 4, 3, 2, and 1 nm) can be seen in Fig. 3.4(c) and Fig. 3.4(d), respectively. It is clear from the figure that the sub-threshold swing and off-state current decrease, while the on-state current increases with the increasing aspect ratio. The effect of fin height in our study is similar to the already reported studies [23-24].

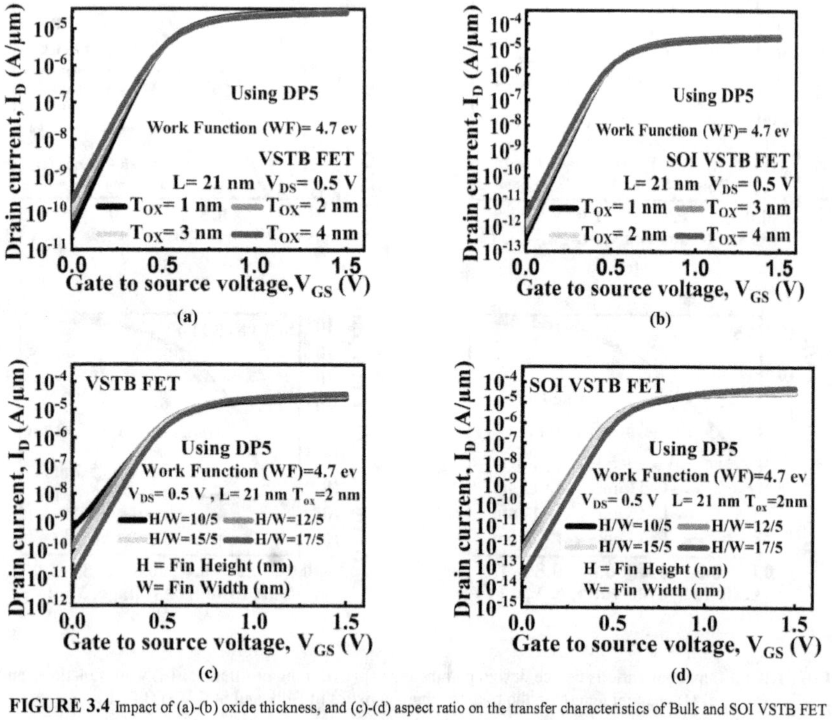

FIGURE 3.4 Impact of (a)-(b) oxide thickness, and (c)-(d) aspect ratio on the transfer characteristics of Bulk and SOI VSTB FET

The I_D-V_{GS} plots of bulk and SOI VSTFET for various aspect ratios are illustrated in Fig. 3.4(c) and Fig. 3.4(d), respectively. Variations of the sub-threshold swing, threshold voltage, on-current, and off-current with various fin heights (H_{fin} = 10, 12, 15, and 17 nm) and fin width (W_{fin} = 5, 4, 3, 2, and 1 nm) can be seen in Fig. 3.4(c) and Fig. 3.4(d), respectively. It is clear from the figure that the sub-threshold swing and off-state current decrease, while the on-state current increases with the increasing aspect ratio. The effect of fin height in our study is similar to the already reported studies [23-24].

3.3.2 Study of different fundamental figures of merit (FoM)

The plots of sub-threshold swing versus channel length and aspect ratio of VSTBFET and SOI VSTB FET are shown in Fig. 3.5(a) and Fig. 3.5(b), respectively. Sub-threshold swing of both the devices improves with an increase in channel length and aspect ratio [22]. It is evident from Fig. 3.5(c) and Fig. 3.5(d) that DIBL decreases as channel length and aspect ratio increase is primarily driven by the enhancement of gate electrostatic control over the channel. In the Vertical Super Thin Body (VSTB) FET and its SOI counterpart, the aspect ratio is increased by increasing the fin height (H_{fin}) or, more significantly, by reducing the fin width (W_{fin}). This geometric change allows the gate to exert a much stronger influence on the potential distribution within the thin silicon body. Regarding the sub-threshold swing shown in Fig. 3.5(b), the physical mechanism involves the gate's ability to fully deplete the channel and quickly transition the device from the "off" state to the "on" state. As the aspect ratio increases, the proximity of the gate to the entire channel volume improves, which effectively suppresses short-channel effects (SCEs) and minimizes leakage current. This results in a sharper sub-threshold slope, bringing the SS closer to the ideal theoretical limit.

In terms of the DIBL reduction shown in Fig. 3.5(d), the increase in aspect ratio serves to shield the source-channel junction from the electric field of the drain. In a device with a high aspect ratio, the gate electrode becomes the dominant source of electrostatic potential, preventing the drain's field lines from penetrating deep into the channel and lowering the potential barrier at the source. This stability is further reinforced by the increase in threshold voltage (V_{th}) that occurs as the aspect ratio rises, which physically makes it more difficult for the drain to influence the barrier height. Furthermore, as noted in the study, the SOI VSTB FET demonstrates superior performance in these metrics compared to the bulk version because the Buried Oxide (BOX) layer provides an additional physical barrier that eliminates substrate leakage and further isolates the source from the drain's influence. This is attributed to the increase in threshold voltage due to an increase in channel length and aspect ratio [22]. It can also be seen from Fig. 3.5(e) that the on-off current ratio increases more significantly with the increase in channel length in the case of SOI VSTBFET as compared with its bulk counterpart [21]. In a nutshell, SOI VSTB FET shows better results than bulk VSTB FET due to the existence of a buried oxide layer which reduces the leakage current by mitigating the source-substrate and drain-substrate p-n junction [21, 22].

3.3.3 Electrical characteristics of BULK VSTB FET and SOI VSTB FET

Fig. 3.6(a) shows the value of I_{ON} and I_{OFF} for the Bulk VSTB FET preferably; the transistors should have a large on-current and a small sub-threshold swing in order to improve the switching speed [25-28]. By considering buried silicon dioxide layer in VSTB-FET structure, the physical drawbacks of VSTB FET, such as reduced carrier mobility due to impurities, increased gate tunneling effect as gate-dielectric thickness reduce, and increased leakage current at p-n junctions has been eliminated [24]. The I_{ON}, I_{OFF}, threshold voltage (V_{TH}), and sub-threshold swing can be determined from Fig. 3.6(b). This figure illustrates the electrical characteristic of 21 nm SOI VSTB FET and compared with the conventional VSTB FET, which indicate that SOI VSTB FET has better electrical characteristics [24-25]. Fig. 3.6(c) depicts the comparative study between the conventional VSTB FET and SOI VSTB FET devices. Silicon on insulator (SOI) VSTB FET has considerably enhanced performance from the aspect of electrical characteristics towards the conventional VSTB FET.

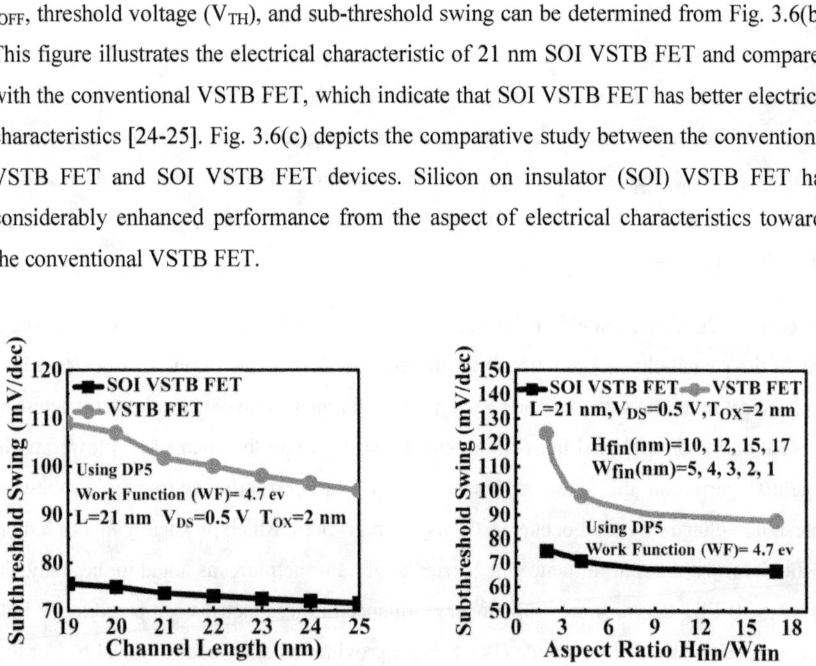

(a) (b)

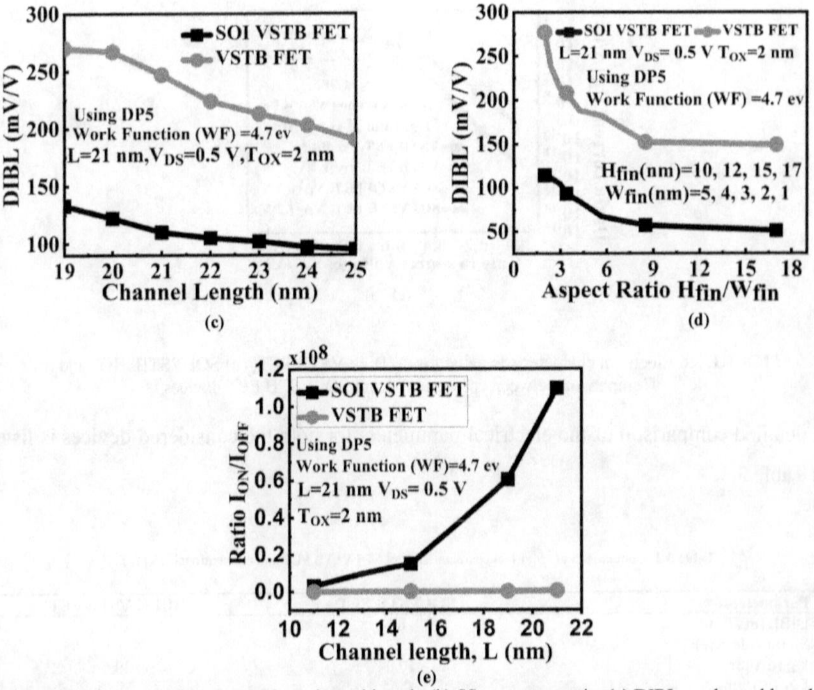

FIGURE 3.5. Various FoMs: (a) SS vs channel length, (b) SS vs. aspect ratio, (c) DIBL vs channel length, and (d) DIBL vs. aspect ratio (e) I_{ON}/I_{OFF} vs. channel length

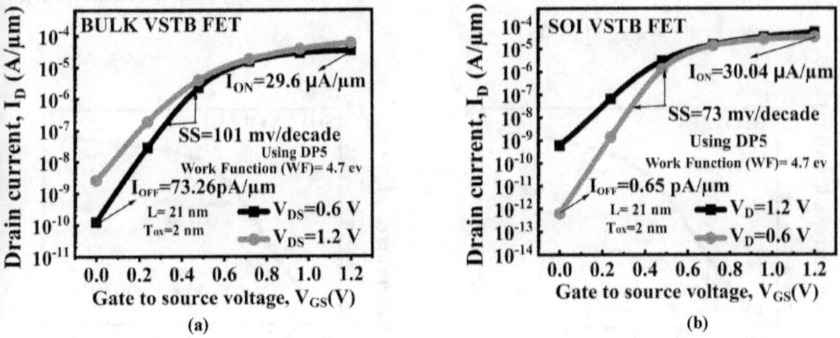

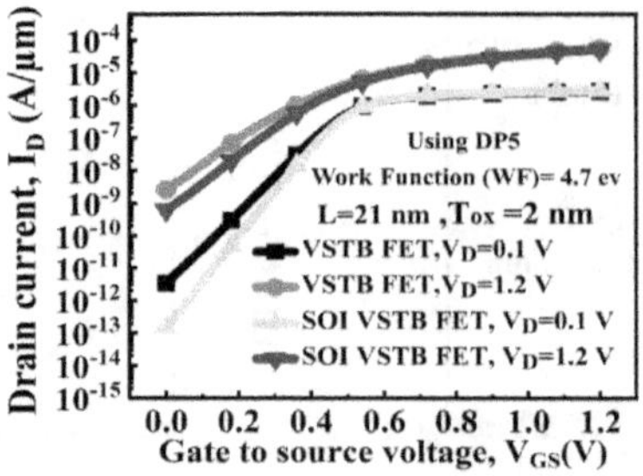

(c)

FIGURE 3.6 Electrical characteristic value for (a) Bulk VSTB FET, (b) SOI VSTB FET, and (c) Comparison between conventional and SOI VSTB FET devices.

A detailed comparison of the electrical parameters for both the considered devices is listed in Table 3.3.

Table 3.3: Comparison of electrical characteristics of SOI VSTB FET and conventional VSTB FET

Parameters	SOI VSTB FET	BULK VSTB FET
DIBL (mV/V)	83	200
SS (mV/decade)	73	101
I_{ON} (µA/µm)	30.04	29.6
I_{OFF} (pA/µm)	0.65	73.26

The I_D-V_{DS} characteristics of bulk VSTB FET device for varying gate voltage are shown in Fig. 3.7(a). Based on the plots, it is seen that the drain current (I_D) proportionately varies with the applied gate voltage [13, 19].

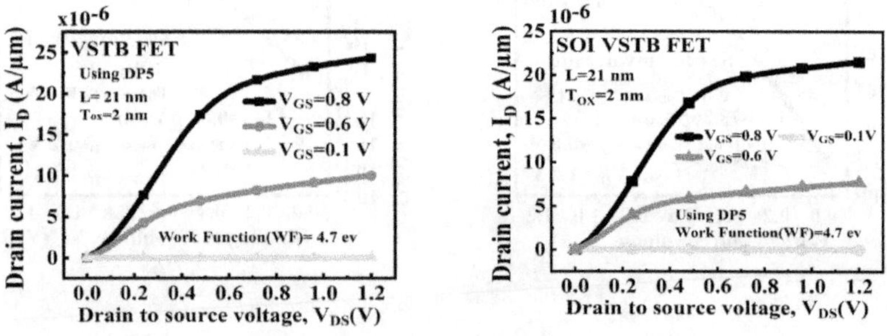

(a)

(b)

FIGURE 3.7 I_D-V_D characteristics for (a) Bulk VSTB FET and (b) SOI VSTB FET.

Fig. 3.7(b) illustrates the performance of the SOI VSTB FET, which is showing the same electrical integrity as bulk VSTB FET with better performance.

3.4 Conclusion

This chapter introduces the Silicon-on-Insulator (SOI)-based (VSTB) FET and provides a comprehensive comparative analysis of its electrical characteristics against those of a Bulk VSTB FET. The findings of this investigation demonstrate that the SOI architecture offers superior electrical performance, outperforming the bulk counterpart across several key metrics. The addition of buried oxide in SOI VSTB FET device is able to lower the sharing of charge carriers between the source and drain region, thereby increasing the ON-state current. Moreover, the leakage current of the SOI VSTB FET is lowered due to the reduction of short channel effects (SCEs). Aditionally, a reduced sub-threshold swing of 73 mV/decade is witnessed in the case of SOI VSTB FET in comparison with the 101 mV/decade for the bulk VSTB FET.

[1] M. Riordan, L. Hodesson, and C. Herring, "Reviews of Modern Physics," *Rev. Mod. Phys.*, vol. 71, pp. S336–S345, 1999.

[2] K. Han, H. Shin, and K. Lee, "Modeling of short-channel effects for deep-submicron SOI MOSFETs," *IEEE Trans. Electron Devices*, vol. 51, pp. 261–269, 2004.

[3] J. C. Tinoco, J. Alvarado, A. G. Martinez-Lopez, and J. A. Martino, "Short-channel effects in multiple-gate SOI MOSFETs," J. Integr. Circuits Syst., vol. 7, pp. 137–144, 2012.

[4] R. T. Bühler, R. Giacomini, M. A. Pavanello, and J. Raskin, "Comparison of analytical and numerical models for short-channel SOI MOSFETs," IEEE Trans. Microw. Theory Techn., vol. 61, pp. 73–76, 2013.

[5] T. Skotnicki, J. A. Hutchby, T. King, H.-S. P. Wong, and F. Boeuf, "The end of CMOS scaling," IEEE Circuits Devices Mag., vol. 21, pp. 16–26, 2005.

[6] F. Balestra, "Silicon-on-Insulator Devices," in Wiley Encyclopedia of Electrical and Electronics Engineering, J. G. Webster, Ed., 2014.

[7] V. Koldiaev and R. Pirogova, "Vertical super-thin body semiconductor on dielectric wall devices and methods of their fabrication," U.S. Patent 8,796,085 B2, 2014.

[8] S. Roy, A. Chatterjee, D. K. Sinha, R. Pirogova, and S. Baishya, "Short channel immunity in vertical MOSFETs," IEEE Trans. Electron Devices, vol. 64, pp. 2106–2112, 2017

[9] Roy Barman, K., Baishya, S. An insight to the performance of vertical super-thin body (VSTB) FET in presence of interface traps and corresponding noise and RF characteristics. *Appl. Phys. A* **125**, 865 (2019). https://doi.org/10.1007/s00339-019-3165-9

[10] X. Chen and C. M. Tan, "Reliability studies of high-k dielectrics in nanoelectronics," Microelectron. Reliab., vol. 54, pp. 1103–1108, 2014.

[11] K. R. Barman and S. Baishya, "Strain engineered vertical super-thin body MOSFET for low power applications," Silicon, vol. 14, pp. 6113–6121, 2022.

[12] Sentaurus Device User Manual, Version J 2019.12 (Synopsis Inc.)

[13] International Technology Roadmap for Semiconductors (ITRS), 2015. [Online]. Available: https://eps.ieee.org/images/files/Roadmap/ITRSFacInt2015.pdf

[14] S. Roy, A. Chatterjee, D. K. Sinha, R. Pirogova, and S. Baishya, "Short channel immunity in vertical MOSFETs," IEEE Trans. Electron Devices, vol. 64, pp. 2106–2112, 2017

[15] X. Chen and C. M. Tan, "Reliability studies of high-k dielectrics in nanoelectronics," Microelectron. Reliab., vol. 54, pp. 1103–1108, 2014.

[16] J.-P. Colinge, C.-W. Lee, A. Afzalian, et al., "Nanowire transistors without junctions," *Nature Nanotechnology*, vol. 5, pp. 225–229, 2010. https://doi.org/10.1038/nnano.2010.15

[17] K. R. Barman and S. Baishya, "An insight to the performance of vertical super-thin body (VSTB) FET in presence of interface traps and corresponding noise and RF characteristics," *Appl. Phys. A*, vol. 125, no. 865, 2019. https://doi.org/10.1007/s00339-019-3165-9
[18] K. R. Barman and S. Baishya, "Subthreshold characteristics of strained vertical super-thin body MOSFETs," Appl. Phys. A, vol. 125, p. 401, 2019.
[19] D. Ranka, A. K. Rana, R. Yadav, K. Yadav, and D. Giri, "Performance analysis of scaled FinFETs," Int. J. VLSI Des. Commun. Syst., vol. 2, pp. 11–22, 2011.
[20] A. Kumar, N. Kar, A. Jaiswal, and A. Kar, "Scaling trends and challenges in FinFET technology," Int. J. Appl. Innovation Eng. Manag., vol. 2, pp. 540–545, 2013
[21] Y. Tsividis and C. McAndrew, *The MOS Transistor*, 3rd ed. Oxford, U.K.: Oxford Univ. Press, 2012
[22] K. E. Kaharudin et al., "Fabrication challenges of FinFET structures," J. Eng. Sci. Technol., vol. 14, pp. 2410–2430, 2019.
[23] M. Charmi, "Impact of mobility degradation on nanoscale FinFETs," Chin. Phys. B, vol. 24, p. 047302, 2015.
[24] E. D. Kurniawan, H. Yang, C. C. Lin, and Y. C. Wu, "Thermal reliability of advanced FinFETs," Microelectron. Reliab., vol. 83, pp. 254–259, 2018.
[25] N. E. I. Boukortt, T. R. Lenka, S. Patanè, and G. Crupi, "Performance optimization of FinFET devices," Electronics, vol. 11, p. 91, 2022.
[26] M. N. I. A. Aziz et al., "Challenges in FinFET technology scaling," J. Telecommun. Electron. Comput. Eng., vol. 6, pp. 45–49, 2014.
[27] K. E. Kaharudin, A. H. Hamidon, and F. Salehuddin, "Impact of scaling trends on FinFET characteristics," Int. J. Eng. Technol., vol. 6, pp. 2070–2079, 2014.
[28] Ferain, I., Colinge, C. & Colinge, JP. Multigate transistors as the future of classical metal–oxide–semiconductor field-effect transistors. *Nature* **479**, 310–316 (2011). https://doi.org/10.1038/nature10676

CHAPTER 4

RF, Analog, and Noise Performance of SOI VSTB FET

4.1 Introduction

In the evolving landscape of semiconductor technology, there is an ever-growing demand for advanced transistor architectures that can efficiently support both high-speed (RF/analog) and low-power applications. While the Vertical Super-Thin Body (VSTB) FET has demonstrated excellent electrical performance, a detailed study on its RF, analog, and noise behavior is essential, particularly with the integration of Silicon-On-Insulator (SOI) technology [1–5]. The addition of the buried oxide (BOX) layer in SOI VSTB FET improves electrostatic integrity by minimizing short channel effects (SCEs) and reducing parasitic capacitances, leading to enhanced signal integrity [5]. However, to evaluate the suitability of SOI VSTB FET for future nanoelectronic applications, it becomes crucial to investigate their analog/RF performance and noise resilience, particularly under the influence of interfacial traps [6]. This chapter presents a systematic comparative analysis of the RF, analog, and noise characteristics of SOI VSTB FET versus Bulk VSTB FET using calibrated 3D Sentaurus TCAD simulations [7-8]. Essential RF figures of merit such as transconductance (g_m), unit-gain cut-off frequency(f_T), and gain-bandwidth product(GBP) are evaluated along with analog metrics like intrinsic gain (g_m/g_d) and transconductance generation factor (TGF). Additionally, the impact of interface traps on noise, including diffusion noise, generation-recombination (G-R) noise, and flicker noise, is analyzed under varying conditions [3]. These investigations confirm that SOI VSTB FET exhibits superior analog/RF performance and better noise immunity than its bulk counterpart [3]. The chapter organization is as follows: Section 4.2 describes the device structure and simulation setup used in this study; Section 4.3 presents and discusses the detailed results; and finally, Section 4.4 summarizes the key findings and conclusions drawn from this work.

4.2 Simulation Set-up

Figures 4.1(a) and 4.1(b) depict three-dimensional schematic representations of the conventional and SOI VSTB FETs, respectively. In addition, Figures 4.1(c) and 4.1(d) show two-dimensional schematic top view and colour codes of the SOI-VSTB FETs, respectively. The procedures for creating such a structure are identical to those mentioned in [6], [9]. The doping concentrations are 10^{19} cm^{-3}, 10^{15} cm^{-3}, 10^{17} cm^{-3}, and 10^{15} cm^{-3} for the source, channel, drain, and substrate, respectively.

Table 4.1: Device dimensions with their corresponding parameter.

Parameters	Dimensions (nm)
t_{body} /Channel thickness	5
t_g	5
t_{ox}	2
L	21
w	4
h	15
h_1	35
h_2	20
h_3	20

The drain is lightly doped compared to source in order to reduce the leakage current (I_{OFF}). To enhance the electrical parameter, a high dielectric HfO_2 (k = 22) with oxide thickness (t_{ox}= 2 nm) and TiN gate are used. We have considered the device dimensions in accordance with the standard ITRS expressed in terms of micrometres [1]. The various key device parameters are listed in Table 4.1 with their corresponding dimensions. Here, effective width (W_{eff}) is equal to the height of the Source/drain/channel (h_1). For both the SOI and bulk devices, the aspect ratio (W_{eff}/L) are same. Three-dimensional Sentaurus TCAD simulation tools were used to mimic the device characteristics [2, 7].

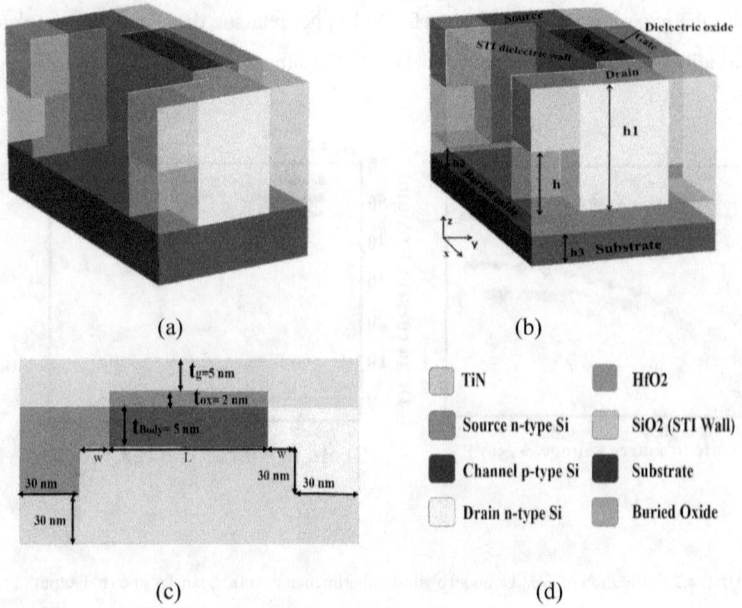

FIGURE 4.1 (a) 3D view of VSTB FET, (b) 3D view of SOI VSTB FET, (c) 2D top view, and (d) Regions name [5].

To analyse the transport of charge carriers, the drift-diffusion model was used [3-4]. The activation of many physics-based models produced genuine effects. Fermi-Dirac statistics was activated to include the effects of high doping concentrations employed in the source/drain area. Models for the influence of carrier mobility on the device doping profile, high electric field, and high-k dielectric oxide [3] were activated, along with the upgraded Lombardi model with high-k deterioration, "doping-dependent Masetti model, high-field saturation model [3]. The velocity saturation model was activated to take into account how a strong electric field affected the carrier velocity [3]. The thin vertical channel may experience stress/strain from a heavy STI dielectric wall [3]. To capture the influence of mechanical stress on carrier transport, the simulation incorporates both stress-induced electron mobility and stress-dependent saturation velocity models [3] to take into account how stress affected the attributes of SOI VSTB FET. Strain effects were also enabled in the simulator by enabling the "deformation potential (DF) model, which also enables flags for strain-induced conduction and valance band-edge changes" [4, 7]. A calibration of TCAD models with the experimental results is performed to get realistic data from the simulator [8]. A nice calibration of the experimental data with the simulator is shown in Fig. 4.2.

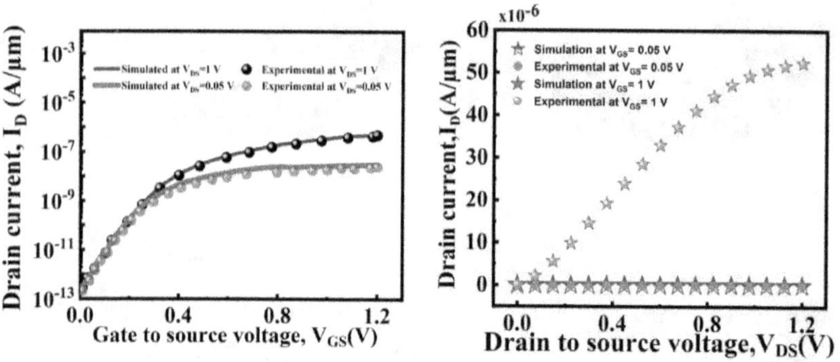

(a) (b)

FIGURE 4.2 Calibration of TCAD model against experimental data (a) Transfer and (b) Output Characteristics [8].

4.3 Results and Discussion

In this article, we have focused on the comparison of AC and Noise parameters for Bulk and SOI devices. Moreover, the comparison of DC parameters are also presented. Some of the crucial characteristics were explored to better understand the Silicon-on-insulator (SOI) and Bulk VSTB FETs: Saturation-current (I_{DSAT}), Threshold voltage (V_{TH}), Input capacitance(C_{gg}), Gate–drain capacitance(C_{gd}), Transconductance (g_m), Gain-bandwidth-product (GBP), Cut-of frequency (f_T), Drain current noise spectral density (S_{vg}ee), Gate voltage electron noise spectral density (S_{vg}ee), and Drain induced barrier lowering (DIBL), On-current (I_{ON}), and Sub-threshold swing. These electrical parameters show how a device behaves in terms of SCE reduction, switching speed, and driving capacity.

4.3.1 Electrostatic potential, electron density, and electron mobility

Fig. 4.3 (a)-(b) shows the Electrostatic potential versus device length i.e., from source to drain. It is seen from the plot that the electrostatic potential of SOI VSTB FET is larger than the Bulk VSTB FET in the channel region due to presence of buried oxide.

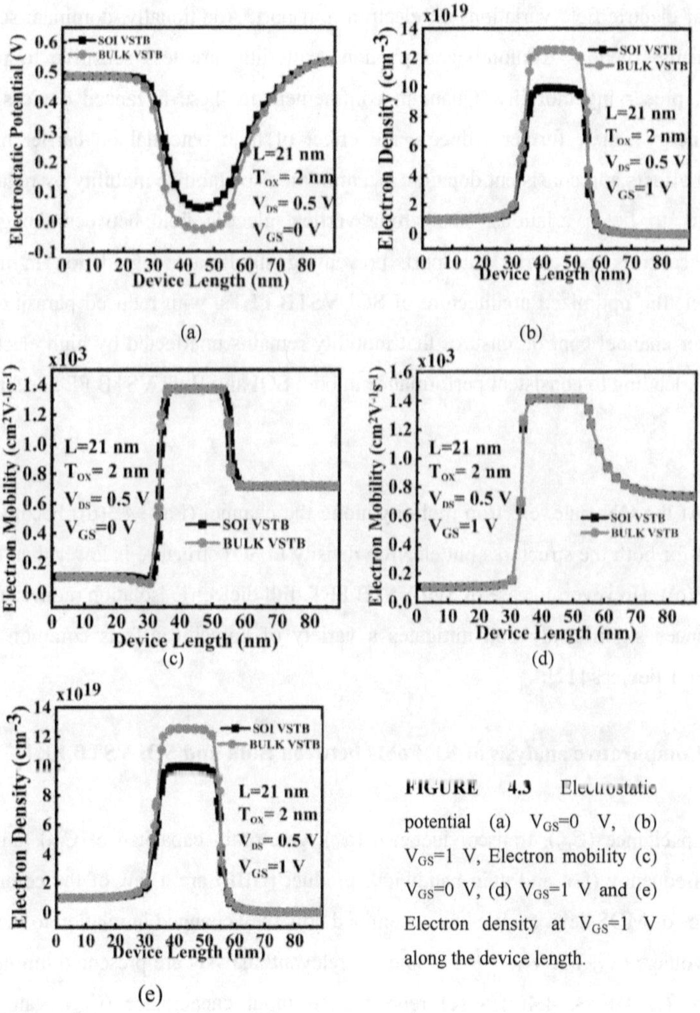

FIGURE 4.3 Electrostatic potential (a) V_{GS}=0 V, (b) V_{GS}=1 V, Electron mobility (c) V_{GS}=0 V, (d) V_{GS}=1 V and (e) Electron density at V_{GS}=1 V along the device length.

The buried oxide layer that is present within the structure of SOI VSTB FETs serves to effectively mitigate the occurrence of SCEs, which are detrimental to device performance, thereby significantly enhancing the overall electrical characteristics of the device by diminishing the potential variations that are induced by these Short Channel Effects, which results in an elevated electrostatic potential when compared to conventional Bulk (VSTB) FETs configurations [10-11]. Electron mobility (Figs. 4.3(c)-(d) in SOI and Bulk VSTB FETs remains constant despite high electrostatic potential in SOI VSTB FETs due to several stabilizing factors. Effective screening by the inversion charge minimizes the impact of electric field variations on electron transport. Additionally, dominant scattering mechanisms, such as Coulomb and phonon scattering, are less sensitive to potential changes, preserving mobility. Quantum confinement in these advanced devices creates stable energy states, further reducing the effect of high potential on carrier mobility. Thermal effects and consistent dopant concentration also stabilize mobility by reducing its sensitivity to field variations. The strong vertical electric field between the gate and channel controls the carrier transport, preventing significant degradation in mobility. Moreover, the optimized architecture of SOI VSTB FETs, with reduced parasitic effects and better channel control, ensures that mobility remains unaffected by high electrostatic potential, leading to consistent performance in both SOI and Bulk VSTB FET technologies [11-15].

At the ON-state, electron mobility along the channel (Figs. 4.3(d)) become nearly constant for both the structures but electron density in SOI structure is lower than the bulk (Fig. 4.3(e)). However, in case of SOI VSTB FET, full dielectric isolation reduces parasitic capacitances significantly and mitigates a variety of parasitic effects common to bulk VSTB FET devices [12].

4.3.1 Comparative analysis of RF FoMs between Bulk and SOI VSTB FET

Input capacitance (C_{gg}), transconductance (g_m), gate-drain capacitance (C_{gd}), unity gain cut-off frequency (f_T), and gain bandwidth product (GBP) are a few of the common RF properties of MOS devices. We have examined how they changed in relation to the gate to source voltage (V_{GS}) at V_{DS} = 0.5 V and the relevant analysis are presented through Figs. 4.4(a) to (f). Figs. 4.4(a) - (c) reported the input capacitance (C_{gg}), gate-to-drain capacitance (C_{gd}), and gate-to-source capacitance (C_{gs}) of SOI (VSTB) FET are diminished relative to Bulk (VSTB) FET due to the buried oxide layer that mitigates parasitic capacitance. Furthermore, SOI (VSTB) FET exhibit reduced junction capacitance owing to minimized junction areas and lower doping concentrations, which alleviate semiconductor material-related capacitance. The fin architecture in SOI (VSTB) FET effectively reduces gate-to-drain and gate-to-source overlap capacitances, while the combination of the buried oxide layer and fin design enhances electrical isolation, further curtailing capacitance.

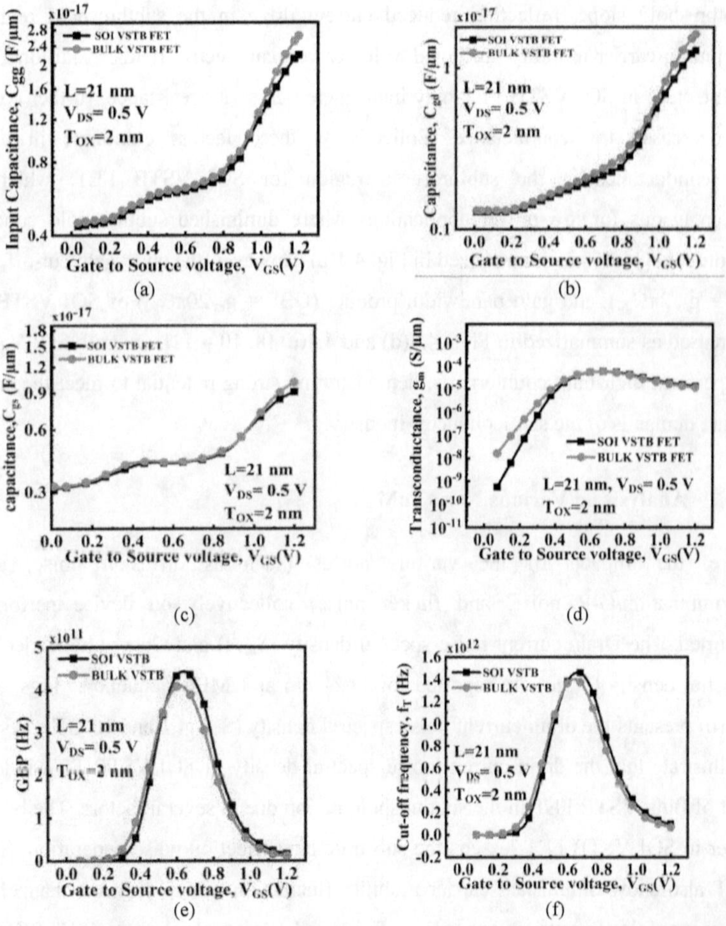

FIGURE 4.4 Various FoMs as a function of gate to source voltage (V_{GS}): (a) Input gate capacitance (C_{gg}), (b) Gate–drain capacitance (C_{gd}), (c) Gate–source capacitance (C_{gs}), (d) Trans-conductance (g_m), (e) Gain-bandwidth-product (GBP), and (f) Cut-of frequency (f_T).

Consequently, SOI VSTB FET demonstrate lower total input capacitance 4.4(a), gate-to-drain capacitance 4.4(b), and gate-to-source capacitance 4.4(c) than Bulk VSTB FET, rendering them more advantageous for high-speed and low-power applications where capacitance reduction is critical. The trans-conductance (g_m) in SOI VSTB FET is diminished in the sub-threshold region relative to Bulk VSTB FET due to multiple factors. The buried oxide layer mitigates the body effect, resulting in diminished transconductance, and displays reduced drain-induced barrier lowering (DIBL) as a consequence of lessened drain voltage impact on the channel. Moreover, SOI VSTB FET generally exhibit a steeper subthreshold slope, reflecting reduced current drive in the subthreshold region and a decline in carrier mobility attributed to lower vertical electric fields. Additionally, the fin architecture in SOI VSTB FET may incur increased series resistance, further contributing to decreased transconductance. Collectively, these factors culminate in diminished transconductance in the subthreshold region for SOI VSTB FET, which can be advantageous for low-power applications where diminished subthreshold conduction is favourable [16-17] as summarized in Fig. 4.4(c). However, the unity gain cut-off frequency ($f_T = g_m/2\pi C_{gg}$), and gain bandwidth product ($GBP = g_m/20\pi C_{gd}$) of SOI VSTB FET are increased as summarized in Figs. 4.4(d) and 4.4(e) [8, 10 – 12]. Overall, SOI VSTB FETs outperform their bulk counterparts, demonstrating strong potential to meet the current and future demands of the semiconductor industry.

4.3.2 Analysis on Various Noise FoM

Here, the impacts of the various noises (such as diffusion noise, generation recombination/G-R noise, and flicker noise) collectively on device performance is reported. The Drain current noise spectral density ($S_{Id}ee$) and Gate voltage electron noise spectral density ($S_{vg}ee$) are reported for L=21 nm at 1 MHz frequency. Figs. 4.5(a) and 4.5(b) presents the drain current noise spectral density (S_{id}) at 1 and 10 GHz, respectively. It illustrate that the drain current noise spectral density of SOI VSTB FET is lower than that of Bulk VSTB FET in the subthreshold region due to several factors. The buried oxide layer in SOI VSTB FET lessens the substrate bias effect on noise generation. SOI VSTB FET also show diminished carrier mobility fluctuation due to reduced channel-substrate interaction, leading to lower noise. Enhanced electrostatic control in SOI structure mitigates random fluctuations affecting drain current. Additionally, SOI VSTB FET has a lower trap density at channel-buried oxide interface, reducing noise from carrier trapping. Finally, lower gate-induced drain leakage in SOI VSTB FET, a major noise source in Bulk VSTB FET contributes to the overall reduction in noise at subthreshold region. These factors collectively account for reduced drain current noise spectral density in SOI VSTB than Bulk VSTB FET [18]. Similarly Figs. 4.5(c) and 4.5 (d) illustrates the gate voltage electron noise spectral density ($S_{vg}ee$) at 1 and 10 GHz, respectively.

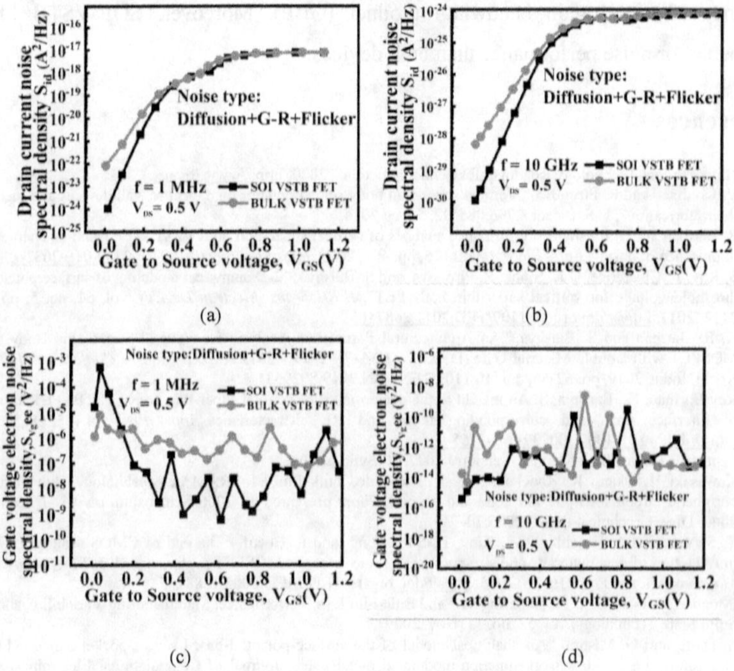

FIGURE 4.5 (a)-(b) Drain current noise spectral density (S_{id}ee) vs. V_{GS} and (c)-(d) Gate voltage electron noise spectral density (S_{vg}ee) vs. V_{GS} for L=21 nm at 1 MHz frequency.

At a given fixed frequency (f = 1 MHz), it is observed from Figs 4.5(c) and 4.5(d) that the values of (S_{vg}ee) for SOI VSTB FET is lower than bulk VSTB FET in sub-threshold region and coincides in saturation region [19].

4.4 Conclusion

In this chapter, a comparison of AC and noise properties for SOI based VSTB FET with Bulk VSTB FET device is investigated. It is seen that the electrical performance of the SOI VSTB FET is excellent. By introducing buried oxide into SOI VSTB FETs, it is possible to increase the ON-state current by reducing charge sharing between the source and drain regions. Additionally, the decrease in short-channel effects(SCEs) lowers the leakage current of SOI VSTB FET. The SOI VSTB FET also has a lower sub-threshold swing of 71.5 mV/decade, whereas, VSTB FET has SS of 99.5 mV/decade. It is also found that RF characteristics of SOI VSTB FETs outperform in terms of transconductance (g_m), cut-off frequency (f_T), and gain bandwidth product (GBP). Moreover, SOI VSTB FET is sustainable for noise performance than bulk devices.

. References

[1] "International Technology Roadmap for Semiconductors," 2020. http://www.itrs.net

[2] V. Koldiaev and R. Pirogova, "Vertical super-thin body semiconductor on dielectric wall devices and methods of their fabrication", U.S. Patent 8,796,085 B2, 5 Aug 2014.

[3] R. Barman and S. Baishya, "Performance analysis of vertical super-thin body (VSTB) FET and its characteristics in presence of noise", *Appl. Phys. A*, vol. **125**, pp. 401, 2019, https://doi.org/10.1007/s00339-019-2682-x

[4] S. Roy, A. Chatterjee, D.K. Sinha, R. Pirogova, and S. Baishya, "2-D analytical modeling of surface potential and threshold voltage for vertical super-thin body FET", *IEEE Trans. Electron Devices*, vol. 64, no. 5, pp. 2106–2112, 2017, https://doi.org/10.1109/TED.2017.2687465.

[5] K. R. Barman and S. Baishya, "An Architectural Parametric Analysis for Vertical Super-Thin Body (VSTB) MOSFET with Double Material Gate (DMG)," *TENCON 2019 - 2019 IEEE Region 10 Conference (TENCON)*, Kochi, India, 2019, pp. 62-66, doi: 10.1109/TENCON.2019.8929531

[6] Roy Barman, K., Baishya, S. An insight to the performance of vertical super-thin body (VSTB) FET in presence of interface traps and corresponding noise and RF characteristics. *Appl. Phys. A* **125**, 865 (2019). https://doi.org/10.1007/s00339-019-3165-9

[7] Sentaurus Device User Manual, Version J 2022.12 (Synopsis Inc.)

[8] Kawasaki H, Okano K, Kaneko A, et al., Embedded bulk FinFET SRAM cell technology with planar FET peripheral circuit for hp32 nm node and beyond. Paper presented at: 2006 Symposium on VLSI Technology, 2006. Digest Technical Paper; 2006:70–71

[9] T. Skotnicki, J.A. Hutchby, T.-J. King, H.-S.P. Wong, and F. Boeuf, "The end of CMOS scaling: toward the introduction of new materials and structural changes to improve MOSFET performance", *IEEE Circuits Devices Mag.*, vol. 21, no. 1, pp. 16–26, 2005. https://doi. org/10.1109/MCD.2005.1388765

[10] Mendez, Horacio, et al. "Comparing SOI and Bulk FinFETs: Performance, Manufacturing Variability, and Cost", Solid State Technology, vol. 52, no. 11, Nov. 2009

[11] S. Dash, and G. Mishra, "An analytical model of the surface-potential-based source-pocket-doped cylindrical-gate tunnel FET with a work-function-modulated metal gate", Journal of Computational Electronics, vol. 19, 2019, 10.1007/s10825-020-01465-x.19

[12] Aziz, M.N.I.A. & Salehuddin, Fauziyah & Mohd Zain, Anis Suhaila & Kaharudin, Khairil Ezwan & Radzi, S. A., "Comparison of electrical characteristics between Bulk MOSFET and Silicon-on-insulator (SOI) MOSFET", *Journal of Telecommunication, Electronic and Computer Engineering*, vol. 6, pp. 45-49, 2014.

[13] Y. Tsividis, C. McAndrew, The MOS Transistor (Oxford University Press, Oxford, 2012)

[14] K. Baruah, R. Das, and S. Baishya, "Impact of trap charge and temperature on DC and Analog/RF performances of hetero structure overlapped PNPN tunnel FET", *Appl. Phys. A*, vol. 126, pp. 856, 2020, https://doi.org/10.1007/s00339-020-04054-8t

[15] Kaharudin, Khairil Ezwan & Salehuddin, Fauziyah & Mohd Zain, Anis Suhaila & Waffle, Frank, "Effect of Channel Length Variation on Analog and RF Performance of Junctionless Double Gate Vertical MOSFET", *Journal of Engineering Science and Technology*, vol. 14, pp. 2410-2430, 2019.

[16] D. Ranka, A. K. Rana, R. K. Yadav, K. Yadav, and D. Giri "Performance evaluation of FD-SOI MOSFETs for different metal gate work function", *arXiv preprint arXiv:*, vol. *1104, pp. 0824,* 2011.

[17] K. R. Barman and S. Baishya, "An Insight into the DC and Analog/RF Response of a Junctionless Vertical Super-Thin Body FET towards High-K Gate Dielectrics", *Silicon*, vol. 14, pp. 6113–6121, 2022. https://doi.org/10.1007/s12633-021-01393-w

[18] E. Simoen, M. G. C. de Andrade, M. Aoulaiche, N. Collaert and C. Claeys, "Low-Frequency-Noise Investigation of n-Channel Bulk FinFETs Developed for One-Transistor Memory Cells," in IEEE Transactions on Electron Devices, vol. 59, no. 5, pp. 1272-1278, May 2012, doi: 10.1109/TED.2012.2186815.]

[19] K. R. Barman and S. Baishya, "Study of enhanced DC and analog/radio frequency performance of a vertical super-thin body FET by high-k gate dielectrics", Int J RF Microw Comput Aided Eng., vol. 32, no. 1, pp. e22940, 2022, doi:10.1002/mmce.2294

CHAPTER 5

Self-Heating Effects in SOI VSTB FET

5.1 Introduction

As semiconductor devices are scaled down to nanometer dimensions, the issue of self-heating effect (SHE) has emerged as one of the key challenges, particularly in Silicon-On-Insulator (SOI) based devices. The buried oxide (BOX) layer in SOI provide excellent electrical isolation, severely limits the dissipation of heat from the active device region to the substrate, causing local temperature rises during device operation [1–4]. In SOI Vertical Super-Thin Body (VSTB) FETs, this effect becomes even more pronounced due to the narrow and vertical structure of the silicon channel, surrounded by STI walls, which further restricts heat flow [5-6]. This thermal accumulation not only degrades the electrical performance reducing drive current (I_{ON}) and increasing leakage but also affects critical RF and analog performance figures such as transconductance (g_m), cut-off frequency (f_T), and gain-bandwidth product (GBP) [7-8]. The reliability of these devices, especially for low-power (LP) and high-performance (HP) applications, depends on understanding and mitigating these thermal effects [9–11]. Therefore, this chapter presents a detailed simulation-based investigation of self-heating effect (SHE) in SOI VSTB FETs is essential for next-generation technologies and specific parameters and behaviors are impacted due to this phenomenon. Using calibrated TCAD simulations, the thermal and electrical characteristics of SOI VSTB FET are examined over a range of temperatures, providing key insights into thermal management strategies [12-13]. The aim of this study is to assess how structural parameters such as body thickness, BOX thickness, and STI dimensions influence the self-heating behavior and consequently, the overall performance of the device [14]. The chapter organization is as follows: Section 5.2 describes the simulation setup, device structure, and modeling strategy used for thermal analysis. Section 5.3 presents and discusses the simulation results on thermal resistance, electrical degradation due to SHE, and its influence on RF/analog characteristics. Section 5.4 concludes the chapter with key findings and recommendations for future work on optimizing thermal performance in SOI VSTB FETs.

5.2 Simulation Set-up

Fig. 5.1(a) illustrate three-dimensional schematic representations of SOI VSTB FET, showcasing the intricate design and structural intricacies of these advanced semiconductor devices. Moreover, Fig. 5.1(c) and 5.1(d) provide a detailed insight into the two-dimensional schematic top view and the intricate color codes employed in the SOI-VSTB FETs, offering a comprehensive visual understanding of the device layout and configuration. The meticulous procedures involved in the fabrication of such a sophisticated structure closely mirror those elucidated in a prior study [15-16], emphasizing the importance of consistency and precision in the manufacturing process to ensure optimal device performance and reliability. Delving into the specifics of the doping concentrations utilized in these FETs, we find values of 10^{19} cm^{-3}, 10^{15} cm^{-3}, 10^{17} cm^{-3}, and 10^{15} cm^{-3} for the source, channel, drain, and substrate regions, respectively, each meticulously tailored to fulfill a specific role in the overall functionality and efficiency of the transistor.

Furthermore, it is noteworthy that the drain region is deliberately subjected to a slightly lower doping concentration compared to the source region, a strategic design choice aimed at mitigating leakage current (I_{OFF}) and optimizing the overall operational characteristics of the device [17]. In a bid to elevate the electrical performance of the FET, a cutting-edge high dielectric material in the form of HfO_2 with a dielectric constant (k) of 22, coupled with a precisely engineered oxide thickness (t_{ox}= 2 nm) and a robust TiN gate material, is employed, underscoring the critical role of advanced materials and fabrication techniques in pushing the boundaries of semiconductor technology [6]. Notably, the dimensional parameters of the device have been meticulously calibrated in strict accordance with the industry-standard International Technology Roadmap for Semiconductors (ITRS) guidelines, which delineate precise specifications in terms of micro-meters, ensuring uniformity and compatibility across different manufacturing processes and technologies [1].

Table 5.1: Device dimensions with their corresponding parameter

Parameters	Dimensions (nm)
t_{body}	5
t_g	5
t_{ox}	2
L	21
W	4
h_1	35
h_2	20
h_3	20
h4	15

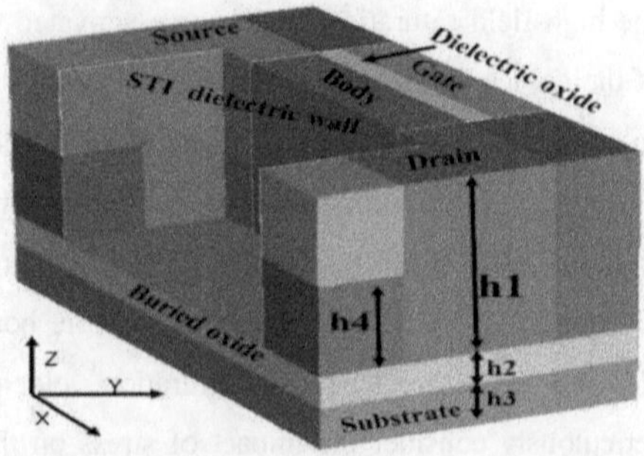

(a)

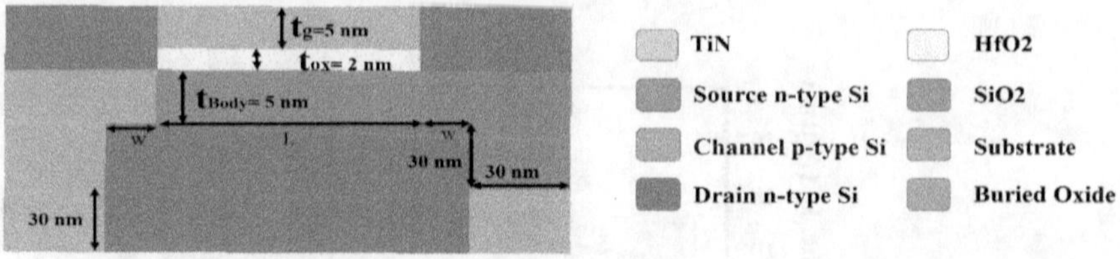

(b) (c)

FIGURE 5.1 (a) 3D view of SOI VSTB FET, (b) 2D top view, and (c) Regions name [6].

For a comprehensive overview of the key operational characteristics and performance metrics of the SOI-VSTB FETs, a detailed tabulation of various crucial device parameters is provided in Table I, presenting a holistic view of the device's intricate design and functional attributes. The utilization of 3D Sentaurus TCAD simulation tools was employed in order to replicate and emulate the device characteristics as described in references [6, 37]. In order to delve into and examine the intricate details of the transport of charge carriers within the device, the drift-diffusion model was utilized as indicated in references [38-40]. The activation and utilization of a multitude of physics-based models yielded authentic and genuine effects that further enhanced the simulation results. Specifically, the incorporation of Fermi-Dirac statistics was pivotal in capturing and encompassing the effects arising from high doping concentrations that are typically utilized in the source/drain region. Furthermore, various models were triggered to account for the impacts of carrier mobility on the device's doping profile, high electric field behaviours, as well as the influence of high-k dielectric oxide, all of which are detailed in reference [39]. Moreover, an advanced version of the Lombardi model, featuring high-k deterioration, the "doping-dependent Masetti model", and the high-field saturation model were activated to provide a comprehensive understanding of the device behavior [39]. The activation of the velocity saturation model was essential to accurately portray how a robust electric field influenced the carrier velocity within the device [39]. Additionally, it was noted that the thin vertical channel within the device might be subjected to stress and strain originating from the thick STI dielectric wall [39]. Consequently, to address this phenomenon, both the stress-induced electron mobility model and the stress-dependent saturation velocity model were brought into play [39] to meticulously consider the impact of stress on the attributes of the SOI VSTB FET.

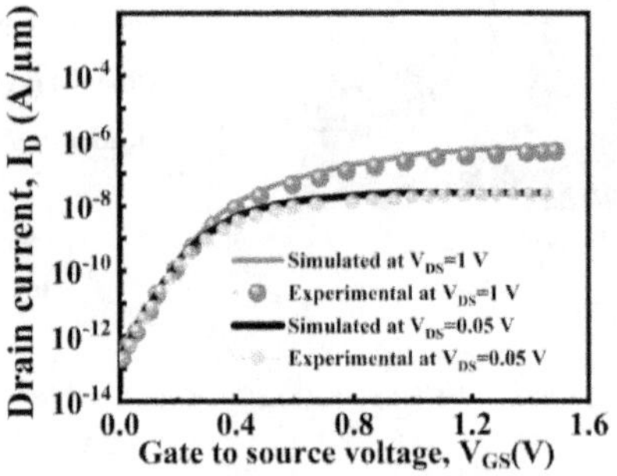

FIGURE 5.2 Calibration of TCAD model against experimental data [41].

Furthermore, the incorporation of strain effects in the simulator was facilitated by enabling the "deformation potential (DF) model, which also encompasses flags for strain-induced conduction and valence band-edge alterations" as detailed in references [37, 39]. The thermodynamic model, as well as hydrodynamic models, the drift-diffusion transport model, and the density gradient model, have all been incorporated into device simulations. These simulations are utilized for the purpose of calculating the energy transport of the carrier within the transport equation. Within the context of this particular research paper, a unique approach is taken in which the electron temperature is computed in a manner that differs from the calculation of the lattice temperature. This differentiation is carried out with the specific aim of investigating the phenomenon known as self-heating effects (SHE). Following these complex modeling techniques, a rigorous calibration of the TCAD models against the experimental results was meticulously carried out to obtain realistic and reliable data from the simulator as referenced in [41]. The harmonious calibration of the experimental data with the simulator outputs is eloquently showcased in Fig. 5.2.

5.3 Results and Discussion

In this chapter, our investigation has been focused on exploring the self-heating effects (SHE) exhibited by 21 nm SOI Vertical Super Thin Body (VSTB) FET. Through our analysis, we have delved into the intricate details surrounding the self-heating effects (SHE) observed in SOI VSTB FETs, considering a wide range of factors such as different gate lengths, VSTB lengths (t_{body}), and variations in oxide thicknesses (t_{ox}). It is noteworthy and intellectually stimulating to observe that the highest temperature within the lattice structure, known as the maximum lattice temperature, as well as the overall amount of heat produced as a result of the absorption of carrier particles traveling at a significantly high velocity, reaches its peak specifically at the drain terminal of the system. This observation sheds light on the intricate dynamics at play within the system, highlighting the critical role of the drain terminal in managing and distributing the thermal energy generated during the reception of carriers moving at high speeds.

Some of the crucial characteristics were explored to better understand the Silicon-on-insulator (SOI) VSTB FETs: Threshold voltage (V_{TH}), Input capacitance (C_{gg}), Gate–drain capacitance (C_{gd}), Trans-conductance (g_m), Gain-bandwidth-product (GBP), Cut-of frequency (f_T), Drain induced barrier lowering (DIBL), On-current (I_{ON}), and Sub-threshold swing. These electrical parameters show how a device behaves in terms of SHE effect.

5.3.1 Thermal analysis of SOI VSTB FET

The thermal properties relating to various thicknesses (t_{body}) of the VSTB FET on Silicon-On-Insulator (SOI) technology with a channel length of 21 nm operating at a voltage of 0.5 volts can be observed in Fig. 5.3(a). The data illustrates that as the temperature rises, there is a decrease in thermal conductivity. Moreover, it shows that with an increase in VSTB thickness, there is a corresponding increase in thermal conductivity. It is evident that as the film thicknesses decreases, the thermal conductivity exhibits a significant and rapid decline. The phenomenon of self-heating effects (SHEs) plays a crucial role in creating a specific high-temperature zone within the device channel near drain. This localized high-temperature region causes the lattice temperature to increase, consequently impacting the transport mechanisms as well as the recombination processes of carriers (electrons and holes) [18].

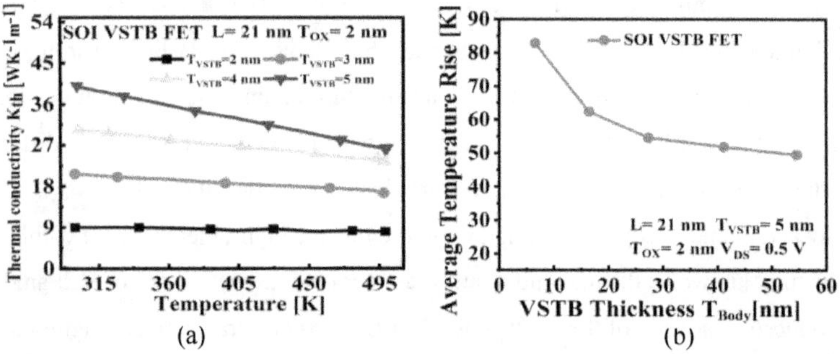

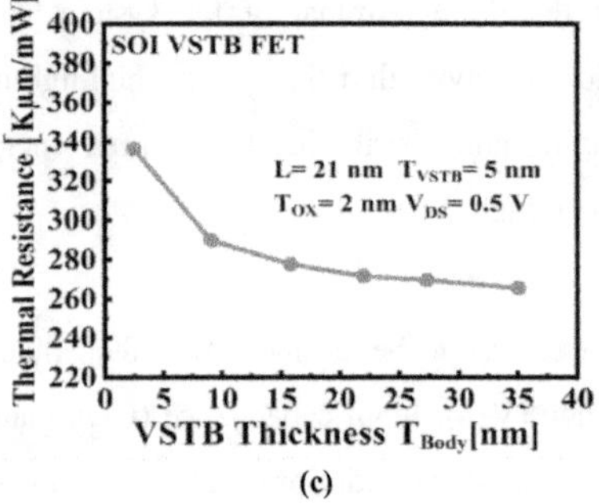

FIGURE 5.3 (a) Thermal conductivity of SOI VSTB FET, (b) Average temperature of SOI VSTB FET, and (c) Thermal resistance of SOI VSTB FET for different VSTB thickness (t_{Body})

In Fig. 5.3(b) and 5.3(c), we can observe that the thermal resistance of the SOI VSTB FET exhibits a decreasing trend as the thickness of the VSTB increases. Additionally, it is evident that the average temperature of the devices also shows a decrease with the increasing thickness of the VSTB.

Table 5.2: Comparison of electrical characteristics of SOI VSTB FET and conventional VSTB FET

Parameters	**300 K**	**350 K**	**400 K**	**450 K**
DIBL (mV/V)	83	93.5	145.6	187.5
SS (mv/decade)	73	81	95	102
I_{ON} (µA/µm)/ I_{OFF}(pA/µm)x10^8	153.4	6.5	0.45	0.001

This trend signifies a relationship between the thermal resistance and the thickness of the VSTB, indicating a potential impact on the overall temperature performance of the devices. [18-19].

5.3.2 DC analysis of SOI VSTB FET

Fig. 5.4(a) illustrates the linear I_D-V_{GS} characteristics observed in a SOI VSTB FET device, providing a visual representation of the changes in gate voltage that affect the device's behaviour.

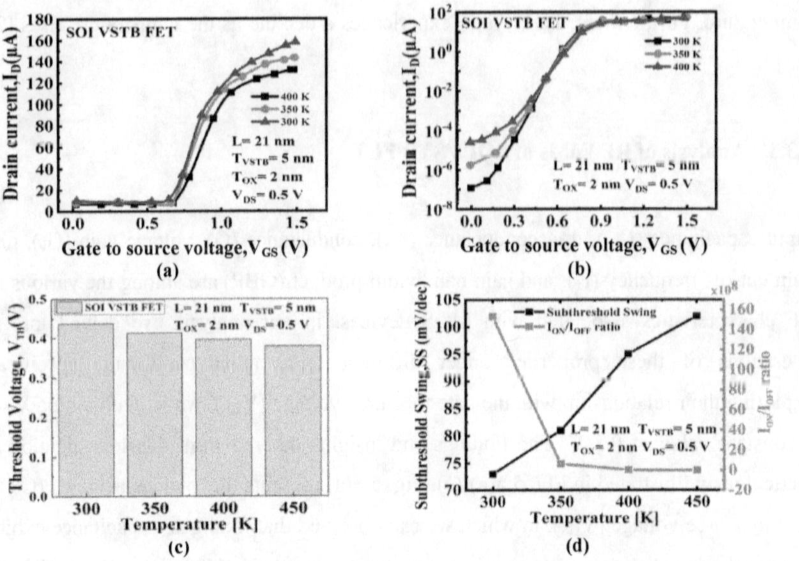

FIGURE 5.4 I_D-V_{DS} characteristics for (a) Linear and (b) Logarithmic SOI VSTB FET, (c) Threshold voltage, and (d) Subthreshold swing and I_{on}/I_{off} ratio for different temperature.

The graphical representation effectively showcases how the drain current (I_D) exhibits a gradual decline as the temperature rises, highlighting the impact of temperature on the device's performance. This particular pattern of decreasing drain current with increasing temperature is also evident in Fig. 5.4(b), where a comparison is made between the device's performance at 300 K and 400 K, showing that at 300 K, the drain current displays superior characteristics compared to the higher temperature of 400 K [20], [21]. These effects are clearly illustrated in Fig. 5.4(c)-(d). The off-state current exhibits a swift increase as a result of the self-heating effect, consequently causing a reduction in the

Ion/Ioff ratio and an escalation in the subthreshold swing, both of which have been meticulously documented in Table II. At off-state, when the device is not actively conducting, a lower bandgap at elevated temperatures plays a crucial role in boosting the quantity of charge carriers present in the channel. Consequently, the drain current (I_D) exhibits a positive temperature coefficient in this scenario. On the other hand, during the on-state when the device is conducting, the mobility of charge carriers tends to deteriorate at higher temperatures due to the scattering effect taking place. As a result, the drain current (I_D) experiences a decline as the temperature rises [21-22]. The threshold voltage of the SOI VSTB FET exhibits a reduction as the temperature rises. Simultaneously, the subthreshold swing of the SOI VSTB FET shows an increment with the increase in temperature. Furthermore, I_{on}/I_{off} ratio experiences a decline as the temperature rises [22], [23].

5.3.3 Analysis of RF FoMs of SOI VSTB FET.

Input capacitance (C_{gg}), transconductance (g_m), conductance (G), voltage gain (G_v), unity gain cut-off frequency (f_T), and gain bandwidth product (GBP) are among the various key RF characteristics associated with MOS devices. In our analysis, we delved into the alterations of these properties under different temperature conditions, specifically exploring their relationship with the gate to source voltage (V_{GS}), while maintaining V_{DS} at a constant value of 0.5 V. The findings and insights derived from this investigation are meticulously illustrated in Fig. 5.5(a)-(e). Fig. 5.5(a) presents the total capacitance (C_{gg}) vs gate to source voltage (V_{GS}), in which we can observed that total gate capacitance exhibits increased when the device operates at elevated temperatures (400 K) in the SOI VSTB FET. This phenomenon is attributed to the impact of self-heating effect (SHE) on the device, which leads to a notable escalation in the lattice temperature of the semiconductor devices [24-26]. In the illustration presented as Fig 5.5(b), a clear observation can be made regarding the trans-conductance behavior of the SOI VSTB FET- it is evident that this trans-conductance decreases with an increase in temperature, a pattern which is linked to the escalation in lattice temperature occurring within the SOI VSTB FET. The reduction in trans-conductance experienced can be predominantly attributed to the phenomenon known as self-heating, a factor that significantly hampers the device's switching performance and subthreshold swing. The impact of self-heating on the device's trans-conductance is direct and profound, ultimately hindering its overall operational efficiency [26-34]. In the

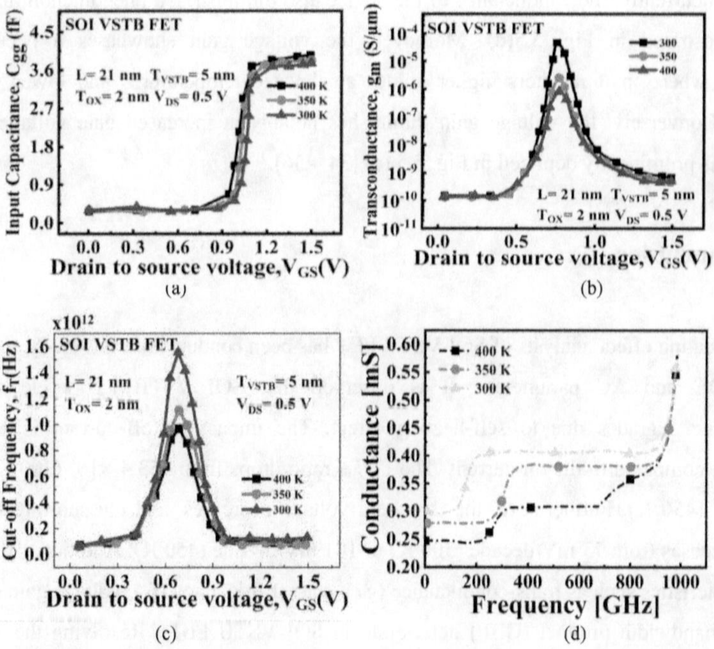

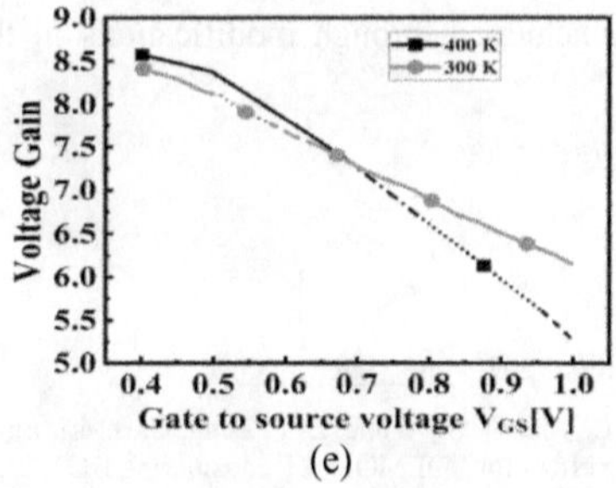

FIGURE 5.5 Various FoMs as a function of gate to source voltage (V_{GS}): (a) Input capacitance (C_{gg}), (b) Trans-conductance (g_m), (c) Cut-of frequency (f_T), (d). Conductance (G), and (e) Voltage gain (G_V)

analysis provided in Fig. 5.5(c), it is evident that the unity gain. The reduction in trans-conductance experienced can be predominantly attributed to the phenomenon known as self-heating, a factor that significantly hampers the device's switching performance and subthreshold swing. The impact of self-heating on the device's trans-conductance is direct and profound, ultimately hindering its overall operational efficiency [26-34]. In the analysis provided in Fig. 5.5(c), it is evident that the unity gain cut-off frequency (f_T =

$g_m/2\pi C_{gg}$) and gain bandwidth product (GBP = $g_m/20\pi C_{gd}$) of SOI VSTB FET exhibit a decreasing trend with the rise in temperature.

Concurrently, the conductance of the device also diminishes, a phenomenon that is clearly illustrated in Fig. 5.5(d). Moreover, the voltage gain showcases a peculiar behaviour where in it registers higher values at elevated temperatures and lower gate voltages. Conversely, the voltage gain diminishes notably at increased gate voltages, a trend that is prominently depicted in Fig. 5.5 (e) [34 – 36].

5.4 Conclusion

The self-heating effect analysis of SOI VSTB FET has been conducted in this study across various DC and AC parameters. It is observed that SOI VSTB FET's electrical performance degrades due to self-heating effect. The impact on off-current is more significant compared to the on-current. The I_{on}/I_{off} ratio drops from 153.4×10^8 (300 K) to 0.001×10^8 (450 K). Furthermore, the threshold voltage decreases, and the sub-threshold swing increases from 73 mV/decade (300 K) to 101 mV/decade (450 K). Additionally, the RF characteristics such as trans-conductance (g_m), cut-off frequency (f_T), voltage gain (G), and gain bandwidth product (GBP) deteriorate in SOI VSTB FETs. Resolving the self-heating effect in these SOI VSTB FETs can be achieved through modifications in the device structure.

References

[1] "International Technology Roadmap for Semiconductors," 2020. http://www.itrs.net

[2] Li, Y.-F., Xu, L. D., Ni, T., Wang, J., Gao, L., Li, D. L., Ma, Q. G., Wang, Z. J., Zeng, C. B., Li, B., & Luo, J. J. (2024). Heat dissipation path of self-heating effect for SOI MOSFET considering BOX layer and TiN barrier layer. https://doi.org/10.1088/1361-6463/ad2006

[3] Ghasemi, H., Mozaffari, M.H. A Simple Proposal to Reduce Self-heating Effect in SOI MOSFETs. *Silicon* **13**, 4189–4198 (2021). https://doi.org/10.1007/s12633-020-00773-y

[4] K. Petrosyants and D. Popov, "Self-Heating Investigation in SOI MOSFET Structures with High Thermal Conductivity Buried Insulator Layers," *2020 36th Semiconductor Thermal Measurement, Modeling & Management Symposium (SEMI-THERM)*, San Jose, CA, USA, 2020, pp. 56-60, doi: 10.23919/SEMI-THERM50369.2020.9142844.

[5] V. Koldiaev and R. Pirogova, "Vertical super-thin body semiconductor on dielectric wall devices and methods of their fabrication", U.S. Patent 8,796,085 B2, 5 Aug 2014.

[6] K. R. Barman and S. Baishya, "An Architectural Parametric Analysis for Vertical Super-Thin Body (VSTB) MOSFET with Double Material Gate (DMG)," *TENCON 2019 - 2019 IEEE Region 10 Conference (TENCON)*, Kochi, India, 2019, pp. 62-66, doi: 10.1109/TENCON.2019.8929531

[7] C. Fiegna, Y. Yang, E. Sangiorgi and A. G. O'Neill, "Analysis of Self-Heating Effects in Ultrathin-Body SOI MOSFETs by Device Simulation," in *IEEE Transactions on Electron Devices*, vol. 55, no. 1, pp. 233-244, Jan. 2008, doi: 10.1109/TED.2007.911354.

[8] Raleva, K., Vasileska, D., Hossain, A., Yoo, S. K., & Goodnick, S. M. (2012). Study of self-heating effects in SOI and conventional MOSFETs with electro-thermal particle-based device simulator. *Journal of Computational Electronics*, *11*(1), 106–117. https://doi.org/10.1007/S10825-012-0384-0

[9] S. Rathore, R. K. Jaisawal, P. N. Kondekar and N. Bagga, "Demonstration of a Nanosheet FET With High Thermal Conductivity Material as Buried Oxide: Mitigation of Self-Heating Effect," in *IEEE Transactions on Electron Devices*, vol. 70, no. 4, pp. 1970-1976, April 2023, doi: 10.1109/TED.2023.3241884.

[10] Ali A. Orouji, Morteza Rahimian,Dual material insulator SOI-LDMOSFET: A novel device for self-heating effect improvement 'Physica E: Low-dimensional Systems and Nanostructures,Vol.44, Issue 1,2011,Pages 333-338,ISSN 1386-9477,https://doi.org/10.1016/j.physe.2011.09.002. (https://www.sciencedirect.com/science/article/pii/S1386947711003249)

[11] Roig, J., Flores, D., Vellvehi, M., Rebollo, J., & Alloza Millán, J. A. (2001). Novel techniques for reducing self-heating effects in silicon-on-insulator power devices. *International Semiconductor Conference*, *2*, 493–496. https://doi.org/10.1109/SMICND.2001.967513

[12] Sentaurus Device User Manual, Version J 2022.12 (Synopsis Inc.)

[13] Kawasaki H, Okano K, Kaneko A, et al., Embedded bulk FinFET SRAM cell technology with planar FET peripheral circuit for hp32 nm node and beyond. Paper presented at: 2006 Symposium on VLSI Technology, 2006. Digest Technical Paper; 2006:70–71

[14] Atamuratov, A. É., Jabbarova, B. O., Khalilloev, M. M., Yusupov, A., & Loureriro, A. G. (2021). Self-heating effect in nanoscale SOI Junctionless FinFET with different geometries. *Spanish Conference on Electron Devices*, 62–65. https://doi.org/10.1109/CDE52135.2021.9455728

[15] S. Roy, A. Chatterjee, D.K. Sinha, R. Pirogova, and S. Baishya, "2-D analytical modeling of surface potential and threshold voltage for vertical super-thin body FET", IEEE Trans. Electron Devices, vol 64, no. 5, pp. 2106–2112, 2017, https://doi.org/10.1109/TED.2017.2687465.

[16] Colinge, Jean-Pierre & Lee, C & Afzalian, Aryan & Dehdashti Akhavan, Nima & Yan, Ran & Razavi, Pedram & O'Neill, Brendan & Blake, Alan & White, Mary & Kelleher, Anne-Marie & Mccarthy, Brendan & Murphy, Richard, "Nanowire transistors without junctions", Nature nanotechnology, 2010, 5. 225-9. 10.1038/nnano.2010.15.

[17] R. Barman and S. Baishya, "Performance analysis of vertical super-thin body (VSTB) FET and its characteristics in presence of noise", *Appl. Phys. A*, vol. **125**, pp. 401, 2019, https://doi.org/10.1007/s00339-019-2682-x

[18] Bhuyan, Muhibul & Mohammedy, Farseem & Khosru, "Doping profile measurement and characterization by scanning capacitance microscope for pocket implanted nano scale n-MOSFET", World Academy of Science, Engineering and Technology, vol. 80, pp. 1156-1163, 2011.

[19] D. Ranka, A. K. Rana, R. K. Yadav, K. Yadav, and D. Giri "Performance evaluation of FD-SOI MOSFETs for different metal gate work function", arXiv preprint arXiv:, vol. 1104, pp. 0824, 2011.

[20] K. R. Barman and S. Baishya, "An Insight into the DC and Analog/RF Response of a Junctionless Vertical Super-Thin Body FET towards High-K Gate Dielectrics", Silicon, vol. 14, pp. 6113–6121, 2022. https://doi.org/10.1007/s12633-021-01393-w

[21] K. R. Barman and S. Baishya, "Study of enhanced DC and analog/radio frequency performance of a vertical super-thin body FET by high-k gate dielectrics", Int J RF Microw Comput Aided Eng., vol. 32, no. 1, pp. e22940, 2022, doi:10.1002/mmce.22940

[22] Mendez, Horacio, et al. "Comparing SOI and Bulk FinFETs: Performance, Manufacturing Variability, and Cost", Solid State Technology, vol. 52, no. 11, Nov. 2009,

[23] S. Dash, and G. Mishra, “An analytical model of the surface-potential-based source-pocket-doped cylindrical-gate tunnel FET with a work-function-modulated metal gate”, Journal of Computational Electronics, vol. 19, 2029, 10.1007/s10825-020-01465-x.19:5

[24] V. M. van Santen, P. R. Genssler, O. Prakash, S. Thomann, J. Henkel and H. Amrouch, "Impact of Self-Heating on Performance, Power and Reliability in FinFET Technology," *2020 25th Asia and South Pacific Design Automation Conference (ASP-DAC)*, Beijing, China, 2020, pp. 68-73, doi: 10.1109/ASP-DAC47756.2020.9045582.

[25] D’Angelo, R., & Der Agopian, P. G. (2013). Comparative study of self-heating effects influence on triple-gate FinFETs fabricated on bulk, SOI and modified substrates. *Symposium on Microelectronics Technology and Devices*, 1–4. https://doi.org/10.1109/SBMICRO.2013.6676140

[26] S. E. Liu *et al.*, "Self-heating effect in FinFETs and its impact on devices reliability characterization," *2014 IEEE International Reliability Physics Symposium*, Waikoloa, HI, USA, 2014, pp. 4A.4.1-4A.4.4, doi: 10.1109/IRPS.2014.6860642.

[27] Wang, S.-C., Su, P., Chen, K.-M., Chen, B.-Y., & Huang, G.-W. (2012). Investigation of Temperature-Dependent High-Frequency Noise Characteristics for Deep-Submicrometer Bulk and SOI MOSFETs. *IEEE Transactions on Electron Devices*, *59*(3), 551–556. https://doi.org/10.1109/TED.2011.2177664

[28] Wang, S.-C., Su, P., Chen, K.-M., Chen, B.-Y., & Huang, G.-W. (2012). Investigation of Temperature-Dependent High-Frequency Noise Characteristics for Deep-Submicrometer Bulk and SOI MOSFETs. *IEEE Transactions on Electron Devices*, *59*(3), 551–556. https://doi.org/10.1109/TED.2011.2177664

[29] Ou, J.-J., Jin, X., Ma, I., Hu, C., & Gray, P. R. (1998). CMOS RF modeling for GHz communication IC’s. *Symposium on VLSI Technology*, 94–95. https://doi.org/10.1109/VLSIT.1998.689213

[30] L. Wang *et al.*, "Impact of Self-Heating on the Statistical Variability in Bulk and SOI FinFETs," in *IEEE Transactions on Electron Devices*, vol. 62, no. 7, pp. 2106-2112, July 2015, doi: 10.1109/TED.2015.2436351. keywords:

[31] H. Jiang, X. Liu, N. Xu, Y. He, G. Du and X. Zhang, "Investigation of Self-Heating Effect on Hot Carrier Degradation in Multiple-Fin SOI FinFETs," in *IEEE Electron Device Letters*, vol. 36, no. 12, pp. 1258-1260, Dec. 2015, doi: 10.1109/LED.2015.2487045.

[32] Raskin, Jean-Pierre & Chung, Tsung & Kilchytska, Valeria & Lederer, D. & Flandre, Denis. (2006). Analog/RF performance of multiple gate SOI devices: Wideband simulations and characterization. Electron Devices, IEEE Transactions on. 53. 1088 - 1095. 10.1109/TED.2006.871876.

[33] Zhang, A.P. & Rowland, L.B. & Kaminsky, E.B. & Tilak, V. & Allen, A.F. & Edward, B.J.. (2003). Self-heating effects in AlGaN/GaN high-power HEMTs. 15 - 16. 10.1109/DRC.2003.1226850.

[34] Hiramoto, Toshiro & Nagumo, Toshiharu & Ohtou, Tetsu & Yokoyama, Kouki. (2007). Device Design of Nanoscale MOSFETs Considering the Suppression of Short Channel Effects and Characteristics Variations. IEICE Transactions on Electronics. E90C. 10.1093/ietele/e90-c.4.836.

[35] Burke, Peter. (2003). An RF Circuit Model for Carbon Nanotubes. Nanotechnology, IEEE Transactions on. 2. 55- 58. 10.1109/TNANO.2003.808503.

[36] Woerlee, Pierre & Knitel, Mathijs & Langevelde, Ronald & Klaassen, D.B.M. & Tiemeijer, L.F. & Scholten, A.J. & Duijnhoven, Adrie. (2001). RF-CMOS performance trends. Electron Devices, IEEE Transactions on. 48. 1776 - 1782. 10.1109/16.936707.

[37] Aziz, M.N.I.A. & Salehuddin, Fauziyah & Mohd Zain, Anis Suhaila & Kaharudin, Khairil Ezwan & Radzi, S. A., “Comparison of electrical characteristics between Bulk MOSFET and Silicon-on-insulator (SOI) MOSFET”, *Journal of Telecommunication, Electronic and Computer Engineering*, vol. 6, pp. 45-49, 2014.

[38] Y. Tsividis, C. McAndrew, The MOS Transistor (Oxford University Press, Oxford, 2012)

[39] K. Baruah, R. Das, and S. Baishya, “Impact of trap charge and temperature on DC and Analog/RF performances of hetero structure overlapped PNPN tunnel FET”, *Appl. Phys. A*, vol. 126, pp. 856, 2020, https://doi.org/10.1007/s00339-020-04054-8t

[40] M. N I A, Aziz & Salehuddin, Fauziyah & Mohd Zain, Anis Suhaila & Kaharudin, Khairil Ezwan, "Electrical characteristics of PMOS bulk mosfet and pmos silicon-on-insulator (SOI) MOSFET device", *ARPN Journal of Engineering and Applied Sciences*, vol. 11, pp. 6315-6318, 2016.

[41] Kaharudin, Khairil Ezwan & Salehuddin, Fauziyah & Mohd Zain, Anis Suhaila & Waffle, Frank, "Effect of Channel Length Variation on Analog and RF Performance of Junctionless Double Gate Vertical MOSFET", *Journal of Engineering Science and Technology*, vol. 14, pp. 2410-2430, 2019.

CHAPTER 6

Stress-Engineered Si/Ge SOI VSTB FETs: A TCAD Exploration of Bandgap and Performance Optimization

6.1 Introduction

As semiconductor devices continue to scale down to meet the growing demand for high-performance and low-power applications, material engineering has emerged as a key approach to enhance device performance. Germanium channels are gaining attention due to their superior carrier mobility over traditional Si, especially in strained architectures [1-7]. However, the trade-offs between improved electrical performance and challenges such as increased leakage current, higher capacitance, and thermal effects require thorough investigation [8-10]. Therefore, this chapter first focuses on comparative study of using silicon and germanium channels in SOI VSTB FETs, particularly emphasizing how mechanical stress influences these devices at the nanoscale as we know that using strained SiGe for pMOSFETs has also enhanced CMOS circuit performance for both static and dynamic operations [11–13]. As a result, SiGe has gained wide attention as a future CMOS technology due to its superior carrier transport properties [14-15]. Compared to Si, SiGe offers much higher mobility for both electrons and holes, overcoming the carrier velocity limit of around 1×10^{7} cm/s in Si at high electric fields [16–18]. Another advantage is that Si and Ge are completely miscible, allowing for the formation of $Si_{1-x}Ge_x$ alloys with Ge content tailored as needed. That's why this chapter also meticulously investigates and delves into the comparative performance metrics associated with a SOI VSTB FET, utilizing both Si and Ge materials, and employing the advanced 3D Senaturus Technology Computer-Aided Design (TCAD) tool for comprehensive analysis. Furthermore, [19-20] it provides a thorough understanding of the underlying cause behind the consequential impact of architectural stress on transfer characteristics of silicon/germanium (Si/Ge) device [21]. Additionally, analogous effects are similarly observed when strain is applied to the devices under consideration, which reinforces the relationship between mechanical stressors and device performance. Lastly, it is noteworthy that, despite Ge relative performance in terms of C_{gs}, C_{gd} and output conductance being lower than that exhibited by Silicon, its higher transconductance effectively reduces intrinsic delay [22-23]. It enhances crucial performance metrics such as intrinsic gain, the unit-gain cut-off frequency, and the GBP, thereby highlighting the complex interplay between material properties and electronic performance. Also in this chapter, we propose a novel $Si_{1-x}Ge_x$ channel n-type SOI VSTB FET with silicon-based source and drain regions. Bandgap engineering is applied to significantly boost electrical performance. Detailed simulations were carried out, focusing on optimizing parameters such as Ge mole fraction, channel length, and oxide thickness. The impact of germanium concentration on threshold voltage (V_T), subthreshold swing (SS), On current (I_{ON}), and the I_{ON}/I_{OFF} ratio have been thoroughly explored. The results show that adding $Si_{1-x}Ge_x$ to the channel improves electrical characteristics, demonstrating the potential of this design for next-generation CMOS technologies.

This chapter is organized as follows: Section 6.2 presents the device structure and simulation methodology adopted for this analysis, including the material composition and stress conditions applied to SOI VSTB FETs including the implementation of $Si_{1-x}Ge_x$ in the channel region. Section 6.3 discusses the key simulation results, emphasizing the comparative study between Si and Ge channel devices, along with the influence of architectural stress on their electrical, analog, and RF parameters and also a comprehensive discussion focusing on how variations in Ge mole fraction, channel dimensions, and gate oxide thickness impact the electrical parameters such as threshold voltage, subthreshold swing, on/off current ratio, and RF performance metrics. Finally, Section 6.4 concludes the chapter by summarizing the important observations drawn from the study and highlights the role of Ge material and mechanical stress in enhancing the performance of SOI VSTB FETs for advanced electronic applications.

6.2 Device Descriptions and Simulation Set-up

Fig. 6.1(a) illustrate three-dimensional schematic of SOI VSTB FET, showcasing the intricate design and structural intricacies of these advanced semiconductor devices. Moreover, Fig. 6.1(b) and 6.1(c) provide a detailed insight into the two-dimensional schematic top view and the intricate colour codes employed in the SOI-VSTB FETs, offering a comprehensive visual understanding of the device layout and configuration. The meticulous procedures involved in the fabrication of such a sophisticated structure closely mirror those elucidated in a prior study [24-25], emphasizing the importance of consistency and precision in the manufacturing process to ensure optimal device performance and reliability. Delving into the specifics of the doping concentrations utilized in these FETs, we find values of 10^{19} cm^{-3}, 10^{15} cm^{-3}, 10^{17} cm^{-3}, and 10^{15} cm^{-3} for the source, channel, drain, and substrate regions, respectively, each meticulously tailored to fulfil a specific role in the overall functionality and efficiency of the transistor. Furthermore, it is noteworthy that the drain region is deliberately subjected to a slightly lower doping concentration compared to the source region, a strategic design choice aimed at mitigating leakage current (I_{OFF}) and optimizing the overall operational characteristics of the device. In a bid to elevate the electrical performance of the FET, a cutting-edge high dielectric material in the form of HfO_2 with a dielectric constant (k) of 22, coupled with a precisely engineered oxide thickness (t_{ox}= 2 nm) and a robust TiN gate material, is employed, underscoring the critical role of advanced materials and fabrication techniques in pushing the boundaries of semiconductor technology. Notably, the dimensional parameters of the device have been meticulously calibrated in strict accordance with the industry-standard International Technology Roadmap for Semiconductors (ITRS) guidelines, which delineate precise specifications in terms of micro-meters, ensuring uniformity and compatibility across different manufacturing processes and technologies. For a comprehensive overview of the key operational characteristics and performance metrics of the SOI-VSTB FETs, a detailed tabulation of various crucial device parameters is provided in Table I, presenting a holistic view of the device's intricate design and functional attributes. The simulation of the proposed $Si_{1-x}Ge_x$ channel SOI VSTB FET is conducted using the 3D Sentaurus TCAD framework, which self-consistently solves the Poisson and drift-diffusion transport equations to replicate the device's electrical characteristics. To accurately capture the physics of the super-thin vertical body and the heterojunction interface, the density gradient model is integrated to account for quantum confinement effects and the physical impacts of varying Ge mole fractions are implemented through a multi-layered coupling of mechanical and semiconductor physics within the Sentaurus TCAD environment as described in references [26]. In order to delve into and examine the intricate details of the transport of charge carriers within the device, the drift-diffusion model was utilized as indicated in references [27-30].

Table 6.1: Device dimensions with their Corresponding parameter.

Parameters	Dimensions (nm)
t_{body}	5
t_g	5
t_{ox}	2
L	21
W	4
h_1	35
h_2	20
h_3	20
h4	15

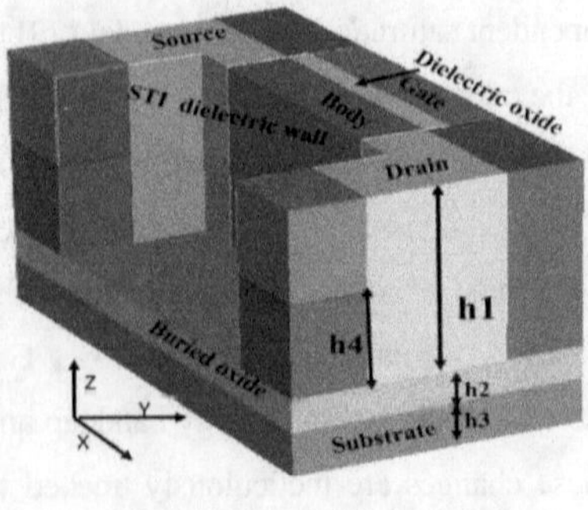

(a)

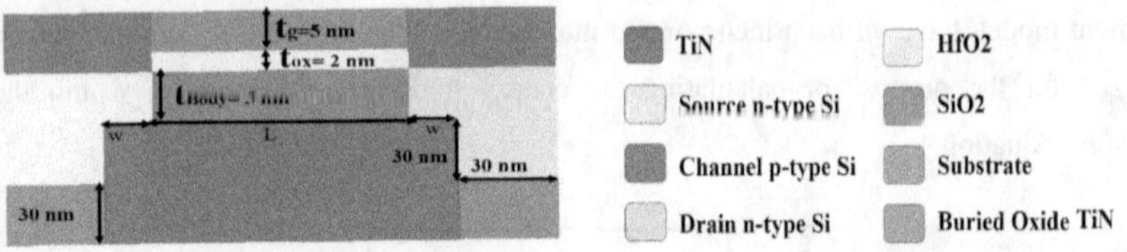

(b) (c)

FIGURE 6.1 (a) 3D view of SOI VSTB FET, (b) 2D top view, and (c) Regions name [14]

The activation and utilization of a multitude of physics-based models yielded authentic and genuine effects that further enhanced the simulation results. Specifically, the incorporation of Fermi-Dirac statistics was pivotal in capturing and encompassing the effects arising from high doping concentrations that are typically utilized in the source/drain region and the degenerate conditions within the $Si_{1-x}Ge_x$ channel. Furthermore, various models were triggered to account for the impacts of carrier mobility on the device's doping profile, high electric field behaviours, as well as the influence of high-k dielectric oxide, all of which are detailed in reference [29]. Moreover, an advanced version of the Lombardi model, featuring high-k deterioration, the "doping-dependent Masetti model", and the high-field saturation model were activated to provide a comprehensive understanding of the device behavior [29]. The activation of the velocity saturation model was essential to accurately portray how a robust electric field influenced the carrier velocity within the device [29]. Additionally, it was noted that the thin vertical channel within the device might be subjected to stress and strain originating from the thick STI dielectric wall . Consequently, to address this phenomenon, The simulation framework implemented both the stress-induced electron mobility model and the stress-dependent saturation velocity model [29] to meticulously consider the impact of stress on the attributes of the SOI VSTB FET. Furthermore, the incorporation of strain effects in the simulator was facilitated by enabling the deformation potential (DF) model, which also encompasses flags for strain-induced conduction and valence band-edge alterations" as the Ge concentration increases, the resulting compressive strain—arising from the lattice mismatch between the Si_{1-x} Ge_x channel and the Silicon buffer leads to a measurable narrowing of the energy bandgap and a corresponding increase in electron affinity. These changes are meticulously tracked to calculate intrinsic carrier concentrations and the subsequent shifts in threshold voltage (V_{th}) as detailed in references [29-30]. The drift-diffusion transport model, and the density gradient model, have all been incorporated into device simulations. These simulations are utilized for the purpose of calculating the energy transport of the carrier within the transport equation.

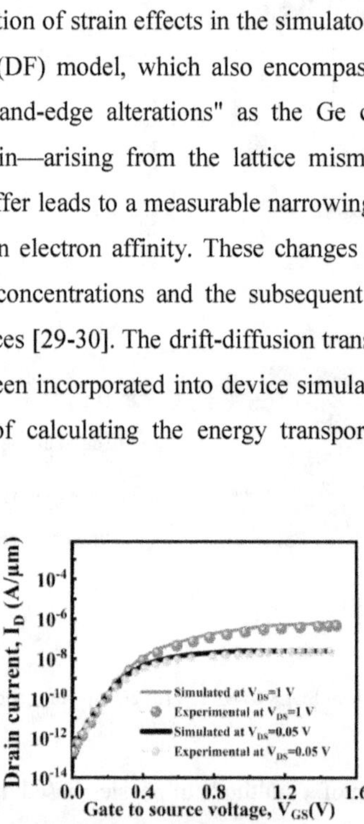

FIGURE 6.2 Calibration of TCAD model against experimental data [19].

Physically, the compressive strain splits the degeneracy of the conduction band valleys, redistributing electrons into valleys with lower effective masses and significantly reducing intervalley scattering rates. These strain-modified parameters incorporated in the drift-diffusion transport model, and the density gradient model. A rigorous calibration of the

TCAD models against the experimental results was meticulously carried out to obtain realistic and reliable data from the simulator as referenced in [31]. The harmonious calibration of the experimental data with the simulator outputs is eloquently showcased in Fig. 6.2.

6.3 Results and Discussion

In this chapter, firstly our investigation has been focused on exploring the impact of the silicon body and germanium body of n-channel of 21 nm SOI Vertical Super Thin Body (VSTB) FET by optimizing the device parameters. SOI (VSTB) FETs with a Germanium body exhibit better I_D-V_{GS} characteristics compared to Si body (Fig. 6.3) due to several advantages. Ge higher electron mobility enables faster switching speeds and better overall performance. Additionally, its lower bandgap energy allows for improved electron flow and reduced leakage current. The subthreshold slope, which measures the transistor's ability to turn on and off, is also improved in Ge-body SOI (VSTB) FETs, leading to reduced power consumption. Furthermore, the SOI (VSTB) FETs structure combined with the SOI platform provides better gate control over the channel, resulting in improved I_D-V_{GS} characteristics. Overall, these advantages make Ge-body I_D-V_{GS} attractive for high-performance, low-power applications, such as high speed analog circuits, low-power digital logic [24]. Now, in the following section, we will delve into the detailed explanation and exploration of the optimization process pertaining to the electrical parameters like On-Current (I_{on}), Off-current (I_{on}), I_{off}/I_{on} ratio, off-state power density, Input capacitance (C_{gg}), Intrinsic Delay, Trans-conductance (g_m), Gain-bandwidth-product (GBP), Cut-of frequency (f_T), and Sub-threshold swing, which play a crucial role in the overall performance of the SOI (VSTB) FET. This optimization is meticulously carried out by strategically varying and adjusting the dimensions of the device, thereby aiming to enhance its efficiency and effectiveness in the simulation.

6.3.1 Impact of Gate Work Function (WF) on Transfer Characteristics of Ge- and Si-Body SOI VSTB FET

The drain current (I_D) versus gate-to-source voltage (V_{GS}) characteristics (Fig. 6.4 (a)-(b)) for SOI VSTB FETs with Si and Ge channels under varying gate work functions (WF =

4.3, 4.5, 4.7 eV) highlight distinct trends driven by material properties and electrostatics. For both Si and Ge bodies,increasing the gate work function leads to rightward shift of the I_D– V_{GS} curve, corresponding to an increase in threshold voltage (V_{GS}) due to enhanced control

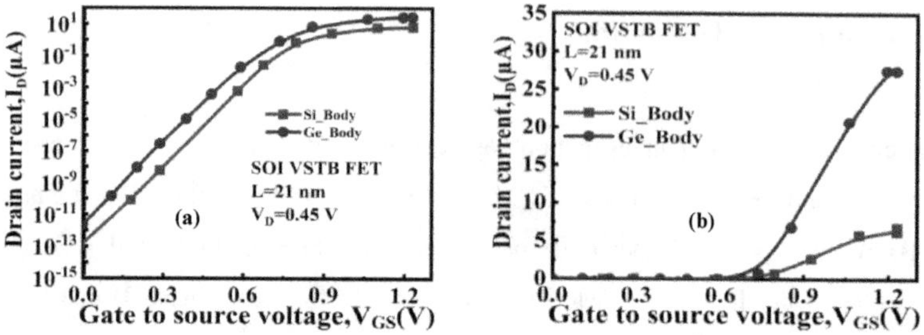

FIGURE 6.3 (a) Logarithmic and (b) Linear I_D-V_{GS} characteristics of SOI VSTB FET for Si and Ge Body.

of the channel potential. This results in Fig. 6.4(a)-(b) show significantly lower off-state leakage current (I_{OFF}), particularly at WF = 4.7 eV. However, the trade-off manifests as a

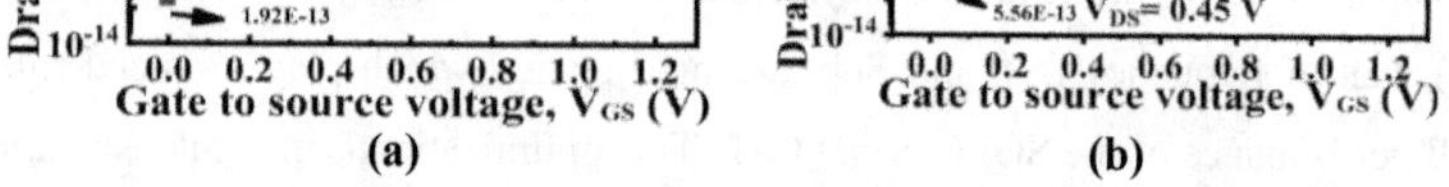

FIGURE 6.4 I_D-V_{GS} characteristics of SOI VSTB FET for (a) Si and (b) Ge Body for different work functions

reduction in on-state current (I_{ON}) with higher WF values. For the Ge-body devices, the I_{ON} remains consistently higher across all WF values compared to the Si body, owing to the superior carrier mobility and lower effective mass of germanium, which enhance carrier transport. Nevertheless, the off-state leakage is also higher in Ge-body FETs due to its smaller bandgap (~0.66 eV) relative to silicon (~1.12 eV), leading to greater thermionic carrier generation. In contrast, Si body devices, while exhibiting lower I_{ON}, maintain better suppression of I_{OFF}, especially at elevated WF [32-33]. This underscores the critical role of gate work function engineering in balancing performance and leakage in SOI VSTB FETs, where Ge channels favor high-performance applications requiring greater drive current, while Si channels remain advantageous in leakage-sensitive, low-power circuits

6.3.2 Impact of Channel Length and Fin Aspect Ratio (H_{Fin}/W_{Fin}) on Si and Ge Body SOI VSTB FET

The on-current of Ge-body SOI VSTB is superior to Si-body SOI VSTB (Fig. 6.5(a) due to

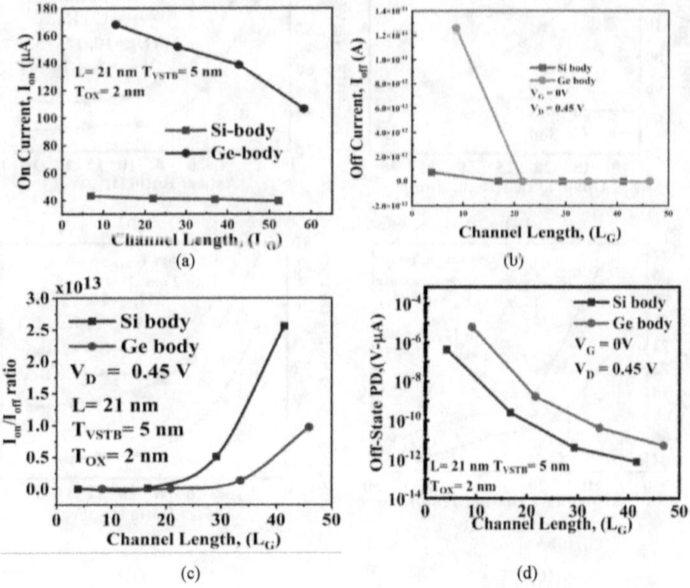

FIGURE 6.5 (a) On-current ,(b) Off-current (c) I_{on}/I_{off} ratio, and (d) off-state power density, for varying channel length for Si and Ge Body

Germanium's higher electron mobility and lower effective mass enable faster electron transport, resulting in increased on-current. Furthermore, improved carrier confinement and reduced interface traps in Ge-body devices contribute to enhanced on-current.

Furthermore, Ge-body SOI VSTB can be strained more easily, providing further mobility and on-current enhancements [34]. The off-current of Ge-body SOI VSTB is higher than Si-body SOI VSTB for short channel lengths (Fig. 6.5(b) due to several factors. Ge narrower bandgap leads to increased leakage current, while its higher dielectric constant results in higher tunneling current, both contributing to elevated off-current. Additionally, Ge-body SOI VSTB FET exhibit a reduced subthreshold slope (Fig. 6.5(c), making it more challenging to turn off the device, and experience higher DIBL (Fig. 6.6(a) and gate-induced drain leakage (GIDL) due to the narrower bandgap. Furthermore, Ge-body devices have higher junction leakage caused by increased diffusion current, further adding to the higher off-current.

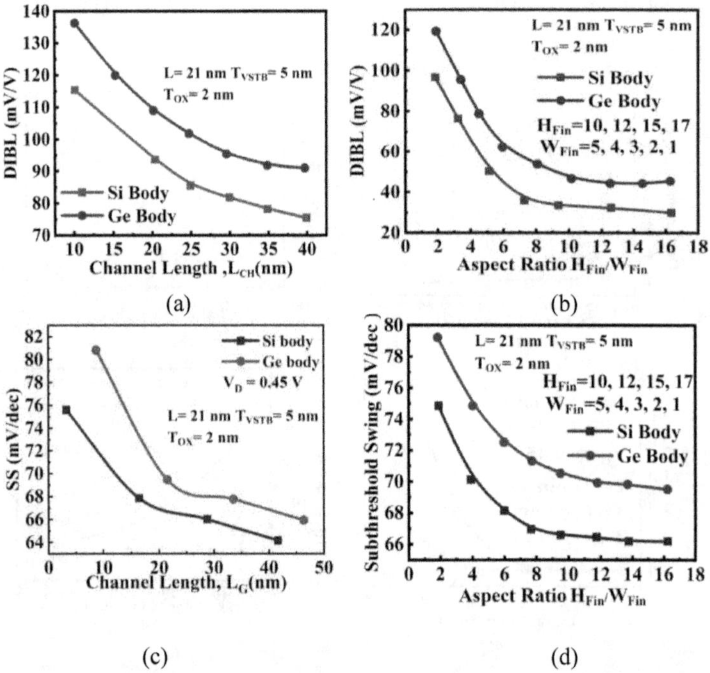

FIGURE 6.6 Comparison of DIBL and Subthreshold Swing for Si and Ge body FETs as a function of (a, c) channel length and (b, d) aspect ratio.

As channel lengths decrease, these factors become more pronounced, resulting in higher off-current in Ge-body SOI VSTB FET compared to Si-body SOI VSTB FET [35]. The off state power density (Fig. 6.5(d)) of Ge-body SOI VSTB FET is higher than Si-body SOI VSTB FET due to increased leakage current and tunneling Current resulting from Ge narrower bandgap and higher dielectric constant. However, as the channel length increases, the off-state power density decreases. This is because longer channels reduce the electric field, leading to decreased tunneling current and leakage, while also improving the subthreshold slope and reducing junction leakage due to lower diffusion current. As a result, the benefits of reduced leakage and tunneling currents outweigh the increased capacitance, leading to decreased off-state power density with increasing channel length. This effect is more pronounced in Ge-body SOI VSTB FET due to their inherently higher leakage and tunneling currents, making channel length optimization crucial for power management in these devices [36-37]. As observed in Fig. 6.6(c)-(d) the SS vs. H_{Fin}/W_{Fin} plot, both Si and Ge body devices show a reduction in subthreshold swing with increasing aspect ratio. This is due to the improved electrostatic gate control over the channel as the fin becomes taller and narrower, effectively enhancing the gate's ability to modulate the channel potential and suppress leakage. Physically, increasing the aspect ratio confines the inversion charge carriers more tightly within the channel, reducing thermionic leakage paths. The Ge body device exhibits a higher SS than its Si counterpart across all aspect ratios. This is attributed to Ge smaller bandgap and higher intrinsic carrier concentration, which makes it more susceptible to leakage currents and worsens the control over off-state conduction compared to silicon. [38-39].

The DIBL vs. aspect ratio plot (Fig 6.6(b)) shows a similar decreasing trend: as the aspect ratio increases, DIBL significantly reduces for both materials. This trend indicates better resistance to drain-induced potential spreading in devices with high aspect ratios. A taller fin creates a stronger vertical field from the gate, which effectively screens the channel from drain potential variations, thus preserving the barrier height at the source end. Again, the Ge body device displays higher DIBL values, owing to its lower conduction band offset and weaker channel-to-drain barrier control. The stronger electric field penetration from the drain in Ge devices reduces the barrier for carrier injection from the source, enhancing DIBL [40].

6.3.3 Comparative Analysis of RF Performance Metrics in Ge- and Si SOI VSTB FET

In Fig. 6.7(a)-(d) and 6.8(a)-(d), the graphs depict the variations of Input capacitance (C_{gg}), Intrinsic Delay, Trans-conductance (g_m),Cut-of frequency (f_T), Gain bandwidth product (GBP), intrinsic gain, and Trans-conductance generation factor (TGF) as a function of gate voltage , while maintaining oxide thickness(t_{ox} =2 nm) and VSTB thickness (t_{Body}= 5 nm). It is evident (Fig: 6.7(a) and 6.8(c)) that the higher mobility and lower effective mass of Ge, resulting in reduced Input capacitance C_{gg} However, the gate-drain capacitance (C_{GD}) increases owing to short-channel effects, where stronger gate influence on the drain enhances overlap capacitance, potentially impacting switching speed.

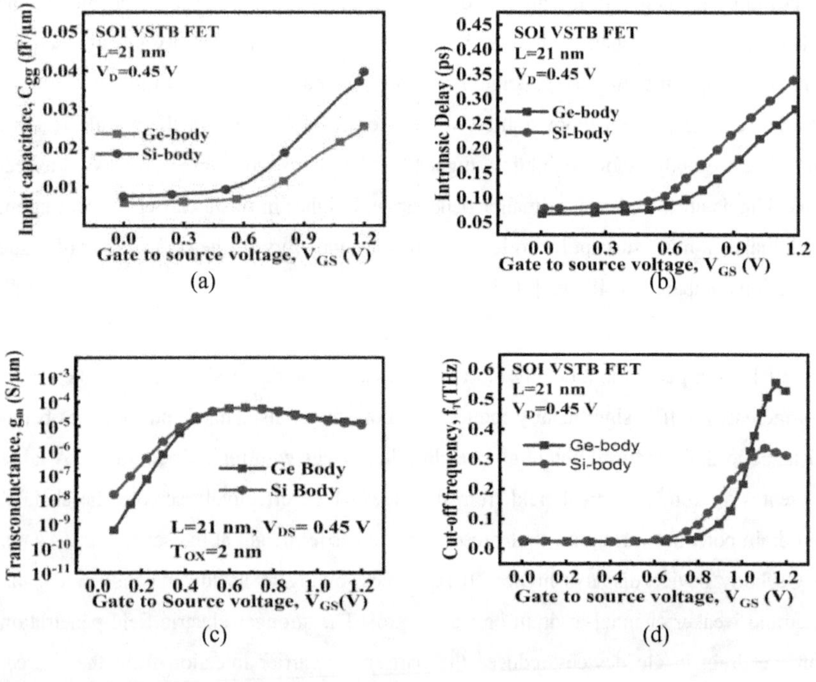

FIGURE 6.7 Various FoMs as a function of gate to source voltage (V_{GS}): (a) Input capacitance (C_{gg}), (b) Intrinsic Delay, (b) Trans-conductance (g_m), (c) Cut-of frequency (f_T), and (d) Gain bandwidth

Additionally, Ge-body VSTB FET typically has thinner gate dielectrics, further reducing input capacitance. Input capacitance increases with increasing channel length because, it increases overlap capacitance: longer channels lead to greater overlap between gate and source/drain regions, increasing capacitance. Fringing capacitance: Longer channels exhibit increased fringing capacitance due to electric field lines extending beyond the gate [41-42]. The lower input capacitance of Ge-body SOI VSTB FET compared to Si-body SOI VSTB FET leads to a lower intrinsic delay (Fig: 6.7(b) and higher intrinsic gain (Fig. 7.8(b), making them suitable for high-speed applications. With reduced input capacitance, Ge-body VSTB FET exhibit decreased RC delay, enabling faster switching times, higher intrinsic gain, and lower intrinsic delay.

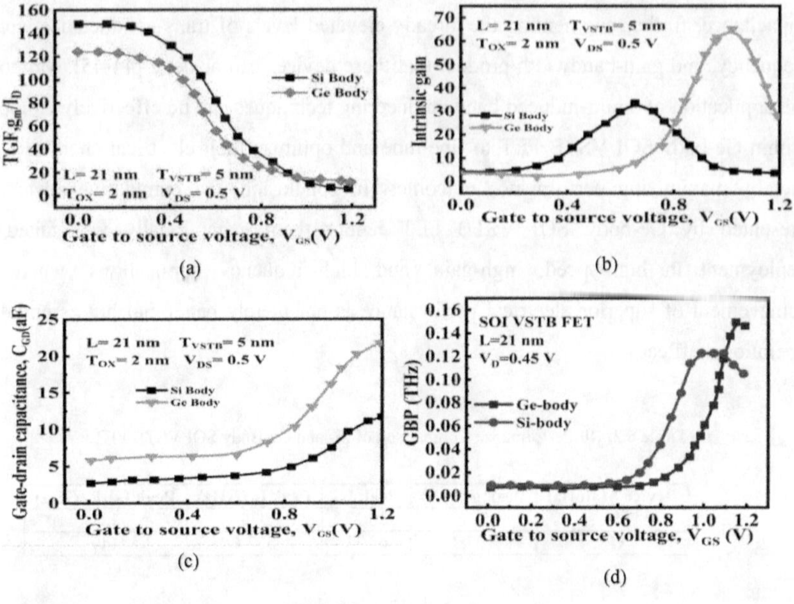

FIGURE 6.8 Additional FoMs as a function of gate to source voltage (V_{GS}): (a) Trans-conductance generation factor (TGF) and (b) Intrinsic Gain, and (c). Gate-drain capacitance (C_{GD}) (d). Gain bandwidth product (GBP)

This means that Ge-body SOI VSTB FET can operate at higher speeds and frequencies compared to Si-body SOI VSTB FET, resulting from their advantageous combination of high mobility, and low capacitance [42-43]. The trans-conductance, cut-off frequency, gain-bandwidth product, and Trans-conductance generation factor (TGF) (Fig. 6.7(c)-(d) and 6.8(a)-(d)) associated with Ge-body SOI VSTB FET. Devices are considerably superior when compared to their Si-body SOI VSTB FET counterparts, a phenomenon that can be attributed to the remarkable intrinsic material properties exhibited by Ge. The enhanced electron mobility characteristic of Ge, in conjunction with its lower effective mass, facilitates a more rapid transport of electrons, which in turn leads to a significant increase in trans-conductance, cut-off frequency, the gain-bandwidth product, , and Transconductance generation factor thereby establishing a distinct advantage over Silicon-based alternatives. Furthermore, Ge-body SOI VSTB FET demonstrate a notable improvement in carrier confinement, which serves to mitigate leakage currents, consequently enhancing their overall electrical performance metrics. In addition, the elevated dielectric constant associated with Ge effectively contributes to an increase in gate capacitance, further augmenting the already elevated levels of trans-conductance, cut-off frequency, and gain-bandwidth product that these devices can achieve [44-45]. Moreover, the application of strain-induced band engineering techniques can be effectively employed within Ge-body SOI VSTB FET to fine-tune and optimize their electrical characteristics, thereby maximizing performance outcomes. In conclusion, the cumulative advantages presented by Ge-body SOI VSTB FET render them exceptionally well-suited for deployment in high-speed, high-gain, and high-frequency applications, where the achievement of superior electrical performance is not merely beneficial but essential for operational efficacy.

Table 6.2. RF Performance Comparison of Si- and Ge-Body SOI VSTB FETs

Device Material	Peak g_m (μS)	Peak IG	Peak f_T (GHz)	Peak GBP (GHz)
Si	57.6	30.23	339	122
Ge	167.3	65.13	557.4	149

6.4 Optimization of Electrical Performance Using $Si_{1-x}Ge_x$ Channel in SOI VSTB FET

In this section, our investigation has been focused on exploring the impact of SiGe for n-channel heterojunction on 21 nm SOI VSTB FET by optimising the device parameters for the different mole fraction (x). Through our analysis, we have delved into the intricate details surrounding the impact of $Si_{1-x}Ge_x$ n-channel heterojunction observed in SOI (VSTB) FETs, considering a wide range of factors such as different gate lengths, VSTB lengths (t_{body}), and variations in oxide thicknesses (t_{ox}).

The illustration that has been provided in Fig 6.9 effectively depicts the changes observed in the energy band diagram due to variation in mole fractions, particularly focusing on the concentration of Ge within Silicon Germanium. Upon careful examination of the diagram, it becomes apparent that as the amount of Germanium in the compound increases, there is a noticeable reduction in the band gap due to the presence of SiGe within the structure of the semiconductor device leads to the generation of uniaxial strain along the Silicon channel. This strain induced by the SiGe component has the effect of enhancing the electron affinity value of the strained-Silicon material. Consequently, this increase in electron affinity brought about by the strain in the material ultimately leads to a reduction in the energy band gap present within the semiconductor device. At the source region, it is observed that as the Ge mole fraction increases, the electron mobility also exhibits an increase.

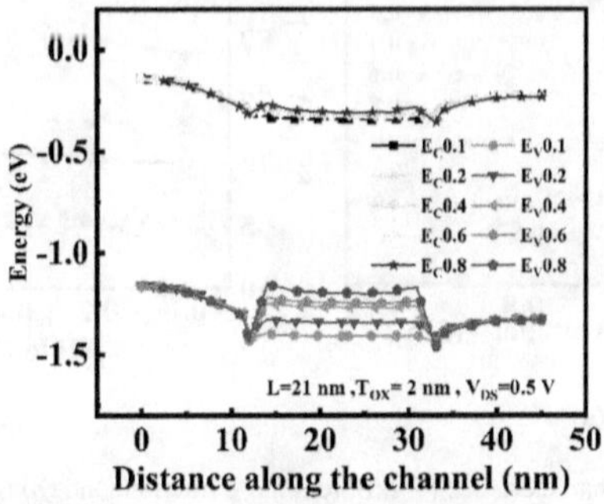

FIGURE 6.9 Effect of different mole fraction on energy band gap

This phenomenon results in the generation of strain on the channel, consequently amplifying the electron mobility within the Si-channel. The enhancement in electron mobility in this scenario is attributed to a reduction in the effective masses of the carrier particles, which occurs as the Ge content in the system rises. [20-21]. Now, in the following section, we will delve into the detailed explanation and exploration of the optimization process pertaining to the electrical parameters like threshold voltage (V_{TH}), input capacitance(C_{gg}), gate–drain capacitance(C_{gd}), transconductance (g_m), gain-bandwidth-product (GBP), cut-of frequency (f_T), I_{ON}/I_{off} ratio, and subthreshold swing, which play a crucial role in the overall performance of the SOI (VSTB) FET. This optimization is meticulously carried out by strategically varing and adjusting the dimensions of the device, thereby aiming to enhance its efficiency and effectiveness in the simulation.

6.4.1 Variation in mole fraction(x) and oxide thickness (t_{ox})

The I_D-V_{GS} characteristics, which depict the relationship between different mole fractions (x) of the VSTB FET implemented on SOI technology, with a channel length measuring 21 nm and operating at a voltage level of 0.5 volts, are visually presented in Fig. 6.10(a) for analysis and examination. Upon careful observation of the data portrayed, it becomes evident that there exists a noticeable reduction in the off-state current as the mole fraction value increases, suggesting a direct correlation between these two parameters.

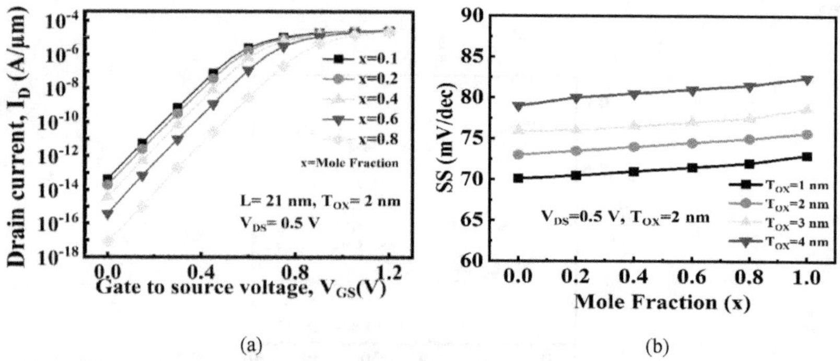

FIGURE 6.10 (a) I_D-V_{GS} characteristics for different mole fraction, and (b) Subthreshold Swing (SS) vs Mole fraction for different oxide thickness..

Furthermore, a compelling trend emerges from the data, revealing that an escalation in the oxide thickness (t_{ox}) results in a proportional augmentation in the subthreshold swing of the SOI (VSTB) FET, as clearly depicted in the graphical representation provided in Fig. 6.10(b), thereby providing valuable insights into the behaviour and performance of the device under varying conditions [46-47].

6.4.2 Variation in channel length (L) of SOI VSTB FET

Figs. 6.11(a)-(c) depict the variations in subthreshold swing, threshold voltage, and I_{on}/ I_{off} ratio as the mole fraction of SiGe changes across different channel lengths, while maintaining oxide thickness(t_{ox} =2 nm) and VSTB thickness (t_{SiGe}= 5 nm). It is evident that the subthreshold swing (SS) experiences significant alterations primarily due to the variations in channel length, with the mole fraction playing a secondary role as shown in Fig. 6.11(a).

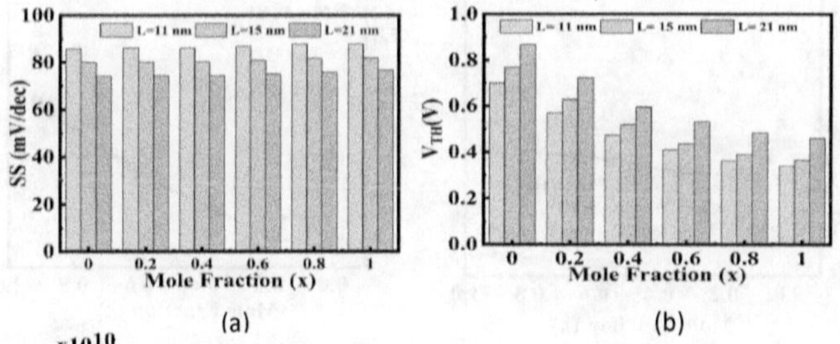

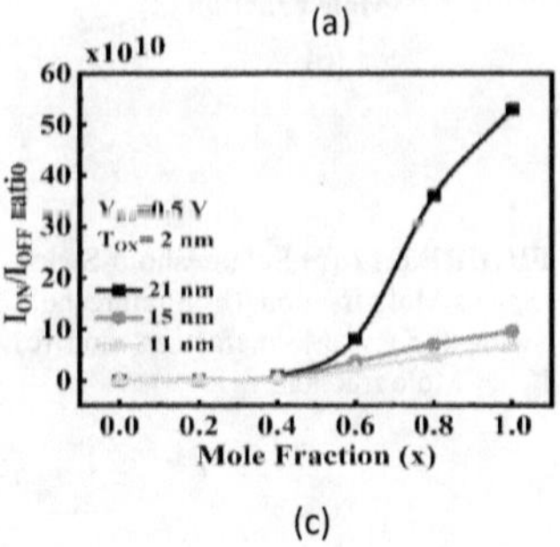

FIGURE 6.11 (a) Subthreshold Swing (SS) vs Mole fraction, (b) Subthreshold Swing (SS) vs Mole fraction, and (c) I_{on}/I_{off} ratio vs Mole fraction for different channel length (L).

Conversely, when examining the threshold voltage (V_{TH}) and I_{on}/I_{off} ratio, it becomes apparent that threshold voltage exhibits downward and I_{on}/I_{off} ratio upward trend with the increase in mole fraction and channel length (L), as illustrated in Fig. 6.11(b) and 6.11(c) [48-49].

6.4.3 Variation in VSTB thickness (t_{SiGe}) of SOI VSTB FET

In Fig. 6.6(a)-(c), the graphical representations illustrate the variations in subthreshold swing, threshold voltage, and I_{on}/I_{off} ratio as the mole fraction of SiGe is modified across It is clearly observable that the subthreshold swing (SS) undergoes significant changes primarily due to the variations in VSTB thickness (t_{SiGe}), where the mole fraction has a secondary impact as depicted in Fig. 6.12(a) [50]. On the contrary, upon closer examination of the threshold voltage (V_{TH}) and I_{on}/I_{off} ratio, it becomes evident that both parameters demonstrate an increasing pattern with the rise in mole fraction; however, the I_{on}/I_{off} ratio reaches its peak at lower VSTB thickness (t_{SiGe} = 5 nm) rather than at higher VSTB thickness (t_{SiGe} = 15 nm) as demonstrated in Fig 6.12(b) and 6.12(c), respectively [51].

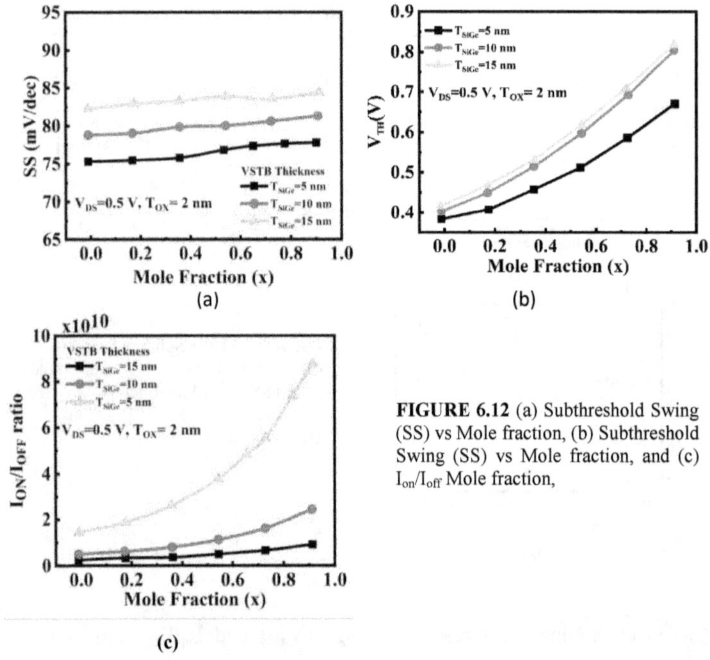

FIGURE 6.12 (a) Subthreshold Swing (SS) vs Mole fraction, (b) Subthreshold Swing (SS) vs Mole fraction, and (c) I_{on}/I_{off} Mole fraction,

6.4.3 Effect of mole fraction on RF FoMs of SOI VSTB FET

Input capacitance (C_{gg}), transconductance (g_m), unity gain cut-off frequency (f_T), and gain bandwidth product (GBP) are the various key RF characteristics associated with MOS devices. In our analysis, we delved into the alterations of these properties under different mole fraction (x) conditions, specifically exploring their relationship with the gate to source voltage (V_{GS}) while maintaining V_{DS} , channel length (L), VSTB thickness (t_{SiGe}), and oxide thickness (t_{ox}) at a constant value of 0.5 V. The findings and insights derived from this investigation are meticulously illustrated in Fig. 6.13(a) to (d).Fig. 6.13(a) presents the total capacitance (C_{gg}) vs gate to source voltage (V_{GS}). The total capacitance of SOI VSTB FET increases with rising mole fraction of Ge in the SiGe hetero-channel SOI VSTB FET due to several factors. Ge higher dielectric constant and carrier mobility boost capacitance, while reduced bandgap, enhances tunneling and capacitance. Additionally, gate-to-channel capacitance, fringing capacitance, and overlap capacitance all increase with higher Ge content, contributing to the overall rise in capacitance. As a result, the incorporation of Ge into the SiGe hetero-channel SOI VSTB

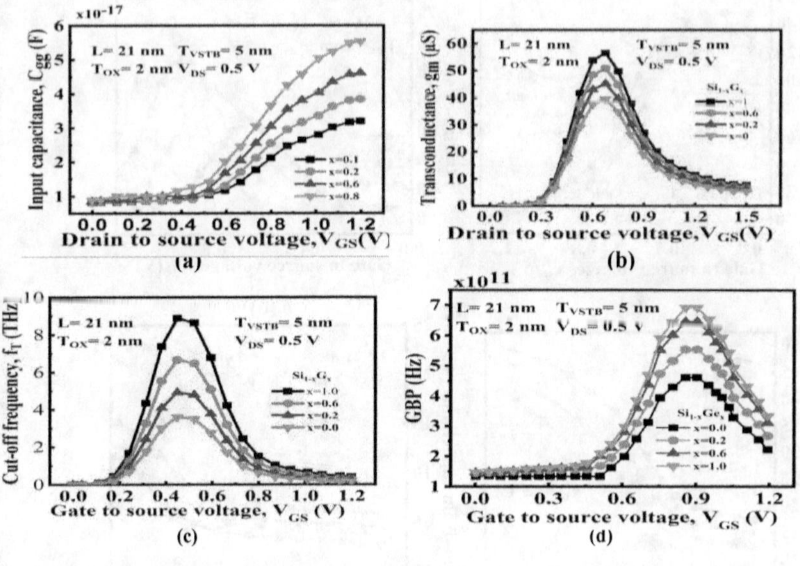

FIGURE 6.13 Various FoMs as a function of gate to source voltage (V_{GS}): (a) Input capacitance (C_{gg}), (b) Trans-conductance (g_m), (c) Cut-of frequency (f_T), and (d) Gain bandwidth product (GBP)

FET leads to increased capacitance, which can impact device performance and power consumption, making it essential to consider these effects in device design and optimization [52]. As presented in Fig 6.13(b), the transconductance of an SOI VSTB FET with a SiGe hetero channel increases with the mole fraction of Ge due to a combination of factors. Ge higher electron mobility enhances channel mobility, while the reduced bandgap of SiGe alloys increases carrier concentration. Additionally, compressive strain from Ge incorporation and quantum confinement effects in the narrow fin width also contribute to the trans-conductance boost. As a result, the overall transconductance of the device increases with higher Ge content, making SiGe hetero channel FinFETs attractive for high-performance applications [53]. The unity gain cut-off frequency ($f_T = g_m/2\pi C_{gg}$) (Fig. 6.13(c)) of a SOI VSTB FET with a SiGe hetero channel increases with the mole fraction of Ge due to reduced effective mass and enhanced channel mobility, allowing for faster carrier transport. However, the gain-bandwidth product ($GBP = g_m/20\pi C_{gd}$) (Fig. 6.13 (d)) paradoxically decreases with increasing Ge content, as higher Ge levels reduce output resistance and increase capacitance,

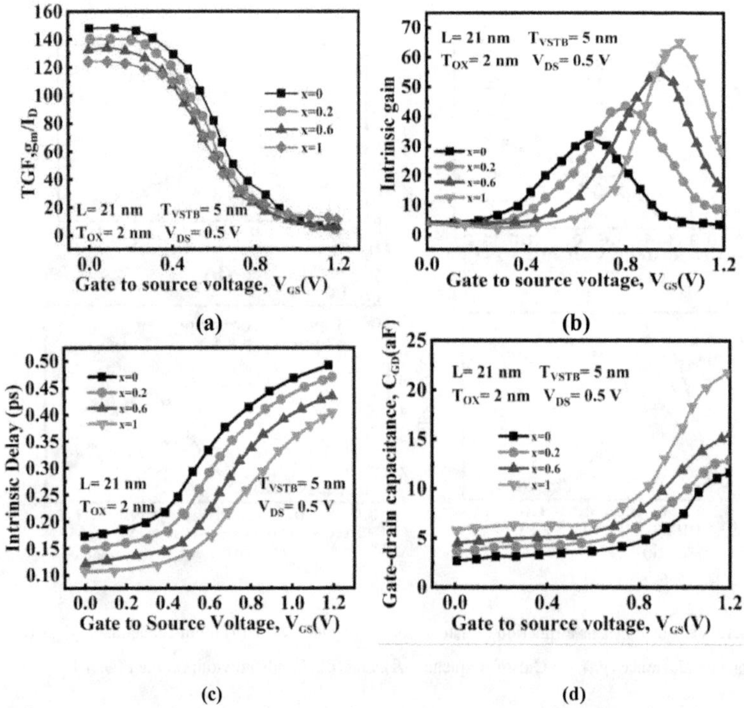

FIGURE 6.14 Various FoMs as a function of gate to source voltage (V_{GS}): (a) Gate- drain capacitance (C_{gd}), (b) TGF (g_m/I_D), (c) Intrinsic delay, and (d) Intrinsic gain

ultimately lowering the GBP. This trade-off highlights the complex interplay between Ge content, device performance, and optimization for specific applications, where a balance between f_T and GBP must be struck to achieve desired outcomes [54]. The variations in intrinsic delay ($\tau = C_{gg}V_D/I_{on}$), intrinsic gain (IG), and transconductance generation factor (TGF), as a function of V_{GS} for different mole fractions are shown in Fig. 6.14(a)–(d), respectively. These parameters were derived from the following equations. Irrespective of the V_{GS} value, the higher mole fraction of Ge in SiGe body demonstrates a significantly lower τ in comparison to lower mole fraction (refer to Fig. 6.14(a). This observation underscores the appropriateness of Ge content in SiGe as a foundational material for high-speed electronic circuits. The influence of different mole fractions of SiGe on IG is observed to exhibit a reversal at both low and high gate biases[55]. At reduced V_{GS}, lower mole fraction achieves a greater IG, whereas at elevated V_{GS}, higher mole fraction matches this performance (see Fig. 6.14(b). This phenomenon can be elucidated by recognizing that IG is directly proportional to transconductance (g_m). Given that g_m is higher for lower mole fraction at lower V_{GS}, the IG of lower mole fraction surpasses that of higher mole fraction. Conversely, Ge demonstrates a higher g_m, and consequently, a superior IG at elevated V_{GS}. Transconductance generation factor (TGF), a pivotal parameter that dictates the efficacy with which the current is employed to achieve a specified transconductance value, reveals exceptionally elevated values within the subthreshold region for different mole fraction of SiGe (illustrated in Fig. 6.14 (a) [56]. This finding indicates that SOI VSTB FET is well-suited for ultra-low power (ULP) applications. Nevertheless, the elevated TGF characteristic of the lower mole fraction renders it marginally more suitable for ULP circuits when compared to its higher mole fraction counterpart. Conversely, TGF presents marginally higher values for the higher Ge content in the device relative to the Si content device in the superthreshold region, a phenomenon attributed to the higher g_m under high gate bias conditions (refer to Fig. 6.13(b)[56].

6.5 Conclusion

This chapter first explores the performance of SOI VSTB FETs using Si and Ge materials. It finds that Ge VSTB FETs outperform Si counterparts, primarily due to higher carrier mobility, which significantly enhances the on-state current (I_{on}) while reducing power consumption. Ge-based devices exhibit higher on-state current and trans-conductance leading to improved intrinsic gain, f_T, and GBP. Although Ge shows slight degradation in off-state current, SS, and higher gate-drain capacitance, it benefits from lower input capacitance and reduced intrinsic delay. Secondly this chapter also investigates the enhancement of electrical parameters in silicon-germanium ($Si_{1-x}Ge_x$) n-channel heterojunction SOI (VSTB) FETs, indicating notable improvements in device efficacy. The findings indicate that higher Ge content decreases the band gap and increases electron mobility due to uniaxial strain for Silicon channel, enhancing charge transport. Additionally, optimizing device parameters like gate lengths and oxide thicknesses is crucial for improving electrical characteristics such as threshold voltage (V_T), I_{on}/I_{off} ratio, and sub-threshold swing(SS). Additionally, the RF characteristics such as input capacitance (C_{gg}), trans-conductance (g_m), unity gain cut-off frequency(f_T), and gain bandwidth product(GBP) have been discussed. The advanced simulation techniques utilized in this research confirm the reliability of the results, facilitating future developments in semiconductor technologies through meticulous manipulation of material properties and device configuration. Ultimately, the incorporation of Si1-xGex channels presents a viable strategy for augmenting the electrical performance of SOI (VSTB) FETs, highlighting their applicability in next-generation electronic technologies.

[1] D. Klaes, J. Moers, A. Tonnesmann, S. Wickenhauser, T. Grabolla, M. Grimm, et al., "Selectively grown vertical Si MOS transistor with reduced overlap capacitances", Thin Solid Film, vol. 336, no. 2, pp. 306-308, 1998.

[2] Y. Liu, K. Ishii, M. Masahara, T. Tsutsumi, H. Takashima, H.Yamauchi, E. Suzuki, "Cross-sectional channel shape dependence of short channel efects in fn-type double-gate metal oxide semiconductor feld efect transistors", Jpn. J. Appl. Phys., vol. 43, pp. 2151, 2004.

[3] Haitao Liu, Zhibin Xiong and J. K. O. Sin, "A novel ultrathin vertical channel NMOSFET with asymmetric fully overlapped LDD," IEEE Electron Device Letters, vol. 24, no. 2, pp. 84-86, Feb. 2003, doi: 10.1109/LED.2002.808153.

[4] Roy Barman, Kuheli & Baishya, Srimanta. (2022). Silicon and Germanium Vertical Super-Thin Body (VSTB) FET: A Comparative Performance Overview Including Architectural Stress-Strain Impact. ECS Journal of Solid State Science and Technology. 11. 10.1149/2162-8777/ac627d

[5] Chemonng Hu. Martin D. McDaniel, Aiting Jiang. Agham Posadas, Alexan-det A Demkov, John G. Ekerdt, and Edward T. Yu, 'A low-leakage epitaxial Inele gate oxide for germanium metal-oxide-semiconductor devices', ACS Appl Mater Interfaces, vol. 8, no. 8, pp. 5416-5423, 2016.

[6] C. II. Lee, L. Y. Yang, J. Heo and G. Yoo, 'Graded Crystalline HfO Gate Di-electric Laver for High-k/Ge MOS Gate Stack,' in IEEE Journal of the Electron Devices Society, vol. 9, pp. 295-299, 2021, doi: 10.1109/JEDS.2021.3058631.

[7] J. Doherty, S. Biswas, E. Galluccio, C. A. Broderick, A. Garcia-Gil, R. Duffy, E. P. O'Reilly, and J. D. Holmes, "Progress on germanium-tin nanoscale alloys", Chem. Mater. vol. 32, no. 11, 4383, 2020.

[8] A. Toriuni and T. Nishimura, 'Germanium CMOS potential from material and process perspectives: Be more positive about germanium', Jpn. J. Appl. Phys., vol. 57, no. 1, 2018.

[9] R. Chau, 'Process and Packaging Innovations for Moore's Law Continuation and Beyond, 2019 IEEE International Electron Devices Meeting (IEDM), 2019, pp. 1.1.1-1.1.6, doi: 10.1109/IEDM19573.2019.8993462.

[10] J. H. Hur and S. H. Jeon, 'III-V compound semiconductors for mass- produced nano-electronics: theoretical studies on mobility degradation by dislocation', Scientific Reports, vol. 6, Feb 2016.

[11] T. Skotnicki, J. A. Hutchby, T. J. King, H. S. P. Wong, and F. Boeuf, "The end of CMOS scaling: toward the introduction of new materials and structural changes to improve MOSFET performance," IEEE Circuits and Devices Magazine, vol. 21, no. 1, pp. 16-26, Jan.-Feb. 2005, doi: 10.1109/MCD.2005.1388765.

[12] K. Uchida, J. Koga, and S. Takagi, "Experimental study on carrier transport mechanisms in double- and single-gate ultrathin-body MOSFETs - Coulomb scattering, volume inversion, and spl delta sub SOI-induced scattering", IEEE International Electron Devices Meeting, pp. 33.5.1-33.5.4, 2003,
[13] L. Silvestri, S. Reggiani, E. Gnani, A. Gnudi and G. Baccarani, "A Low-Field Mobility Model for Bulk, Ultrathin Body SOI and Double-Gate n-MOSFETs With Different Surface and Channel Orientations—Part II: Ultrathin Silicon Films," IEEE Transactions on Electron Devices, vol. 57, no. 7, pp. 1575-1582, July 2010, doi: 10.1109/TED.2010.2049211
[14] O. Weber and S. Takagi, "Experimental examination and physical understanding of the Coulomb scattering mobility in strained-Si nMOSFETs", IEEE Trans. Electron Devices, vol. 55, no. 9, pp. 2386-2396, Sep. 2008.
[15] S. Reggiani, E. Gnani, A. Gnudi, M. Rudan and G. Baccarani, "Low-field electron mobility model for ultra-thin-body SOI and double-gate MOSFETs with extremely small silicon thicknesses", IEEE Trans. Electron Devices, vol. 54, no. 9, pp. 2204-2212, Sep. 2007.
[16] Francis et al. "Double-gate silicon-on-insulator transistor with volume inversion: A new device with greatly enhanced performance." IEEE Electron Device Letters, vol. 8, pp. 410-412, 1987.
[17] Y. Omura, H. Konishi, and K. Yoshimoto, "Impact of fin aspect ratio on short-channel control and drivability of multiple-gate SOI MOSFETs", J. Semicond. Technol. Sci., vol. 8, no. 4, pp. 302–310, 2008.
[18] Y. Liu, K. Ishii, M. Masahara, T. Tsutsumi, H. Takashima, H.Yamauchi, and E. Suzuki, "Cross-sectional channel shape dependence of short channel efects in fn-type double-gate metal oxide semiconductor feld efect transistors", Jpn. J. Appl. Phys., vol. 43, pp. 2151, 2004
[19] Sentaurus Device User Manual, Version J 2022.12 (Synopsis Inc.)
[20] K. R. Barman and S. Baishya, "Study of enhanced DC and analog/radio frequency performance of a vertical super-thin body FET by high-k gate dielectrics", Int J RF Microw Comput Aided Eng., vol. 32, no. 1, pp. e22940, 2022, doi:10.1002/mmce.22940
[21] Roy Barman, Kuheli & Baishya, Srimanta. (2022). Silicon and Germanium Vertical Super-Thin Body (VSTB) FET: A Comparative Performance Overview Including Architectural Stress-Strain Impact. ECS Journal of Solid State Science and Technology. 11. 10.1149/2162-8777/ac627d
[22] F. Balestra, Silicon-on-Insulator Devices, Wiley Encyclopedia of Electrical andElectronics Engineering, edited by John G. Webster (Wiley, 2014).
[23] V. Koldiaev and R. Pirogova, "Vertical super-thin body semiconductor on dielectric wall devices and methods of their fabrication", U.S. Patent 8,796,085 B2, 5 Aug 2014.
[24] S. Roy, A. Chatterjee, D.K. Sinha, R. Pirogova, and S. Baishya, "2-D analytical modeling of surface potential and threshold voltage for vertical super-thin body FET", IEEE Trans. Electron Devices, vol. 64, no. 5, pp. 2106–2112, 2017, https://doi.org/10.1109/TED.2017.2687465.
[25] Colinge, Jean-Pierre & Lee, C & Afzalian, Aryan & Dehdashti Akhavan, Nima & Yan, Ran & Razavi, Pedram & O'Neill, Brendan & Blake, Alan & White, Mary & Kelleher, Anne-Marie & Mccarthy, Brendan & Murphy, Richard, "Nanowire transistors without junctions", Nature nanotechnology, 2010, 5. 225-9. 10.1038/nnano.2010.15.
[26] Aziz, M.N.I.A. & Salehuddin, Fauziyah & Mohd Zain, Anis Suhaila & Kaharudin, Khairil Ezwan & Radzi, S. A., "Comparison of electrical characteristics between Bulk MOSFET and Silicon-on-insulator (SOI) MOSFET", Journal of Telecommunication, Electronic and Computer Engineering, vol. 6, pp. 45-49, 2014.
[27] "International Technology Roadmap for Semiconductors," 2020. http://www.itrs.net
[28] Y. Tsividis, C. McAndrew, The MOS Transistor (Oxford University Press, Oxford, 2012)
[29] K. Baruah, R. Das, and S. Baishya, "Impact of trap charge and temperature on DC and Analog/RF performances of hetero structure overlapped PNPN tunnel FET", *Appl. Phys. A*, vol. 126, pp. 856, 2020, https://doi.org/10.1007/s00339-020-04054-8t
[30] M. N I A, Aziz & Salehuddin, Fauziyah & Mohd Zain, Anis Suhaila & Kaharudin, Khairil Ezwan, "Electrical characteristics of PMOS bulk mosfet and pmos silicon-on-insulator (SOI) MOSFET device", *ARPN Journal of Engineering and Applied Sciences*, vol. 11, pp. 6315-6318, 2016
[31] Kawasaki H, Okano K, Kaneko A, et al., Embedded bulk FinFET SRAM cell technology with planar FET peripheral circuit for hp32 nm node and beyond. Paper presented at: 2006 Symposium on VLSI Technology, 2006. Digest Technical Paper; 2006:70–71.
[32] Kaharudin, Khairil Ezwan & Salehuddin, Fauziyah & Mohd Zain, Anis Suhaila & Waffle, Frank, "Effect of Channel Length Variation on Analog and RF Performance of Junctionless Double Gate Vertical MOSFET", Journal of Engineering Science and Technology, vol. 14, pp. 2410-2430, 2019.

[33] M. N I A, Aziz & Salehuddin, Fauziyah & Mohd Zain, Anis Suhaila & Kaharudin, Khairil Ezwan, "Electrical characteristics of PMOS bulk mosfet and pmos silicon-on-insulator (SOI) MOSFET device", ARPN Journal of Engineering and Applied Sciences, vol. 11, pp. 6315-6318, 2016.

[34] Jungwoo Oh et al., "Additive mobility enhancement and off-state current reduction in SiGe channel pMOSFETs with optimized Si Cap and high-k metal gate stacks," 2009 International Symposium on VLSI Technology, Systems, and Applications, Hsinchu, Taiwan, 2009, pp. 22-23, doi: 10.1109/VTSA.2009.5159274

[35] X. Yu, J. Kang, M. Takenaka and S. Takagi, "Evaluation of Mobility Degradation Factors and Performance Improvement of Ultrathin-Body Germanium-on-Insulator MOSFETs by GOI Thinning Using Plasma Oxidation," in IEEE Transactions on Electron Devices, vol. 64, no. 4, pp. 1418-1425, April 2017, doi: 10.1109/TED.2017.2662217

[36] D. Yadav and D. R. Nair, "Impact of Source to Drain Tunneling on the Ballistic Performance of Si, Ge, GaSb, and GeSn Nanowire p-MOSFETs," in IEEE Journal of the Electron Devices Society, vol. 8, pp. 308-315, 2020, doi: 10.1109/JEDS.2020.2980633.

[37] B. Ho, N. Xu and T. -J. K. Liu, "Study of High-Performance Ge pMOSFET Scaling Accounting for Direct Source-to-Drain Tunneling," in IEEE Transactions on Electron Devices, vol. 58, no. 9, pp. 2895-2902, Sept. 2011, doi: 10.1109/TED.2011.2159008

[38] Yu, E., Lee, WJ., Jung, J. et al. SiGe Heterojunction FinFET Towards Tera-Hertz Applications. J. Korean Phys. Soc. 72, 527–532 (2018). https://doi.org/10.3938/jkps.72.527

[39] Krishna Saraswat, Chi On Chui, Tejas Krishnamohan, Donghyun Kim, Ammar Nayfeh, Abhijit Pethe, High performance germanium MOSFETs, Materials Science and Engineering: B, Volume 135, Issue 3, 2006, Pages 242-249, ISSN 0921-5107, https://doi.org/10.1016/j.mseb.2006.08.014

[40] Vita Pi-Ho Hu and Pin-Chieh Chiu 2018 Jpn. J. Appl. Phys. 57 04FD02 DOI 10.7567/JJAP.57.04FD02

[41] D. Granberry, D. D. Martin, E. Orris and C. M. Chang, "A germanium ultra-high speed switching transistor," in IRE Transactions on Electron Devices, vol. 9, no. 1, pp. 110-110, Jan. 1962, doi: 10.1109/T-ED.1962.14905

[42] G. Hellings et al., "High Performance 70-nm Germanium pMOSFETs With Boron LDD Implants," in IEEE Electron Device Letters, vol. 30, no. 1, pp. 88-90, Jan. 2009, doi: 10.1109/LED.2008.2008824

[43] S. J. Koester, R. Hammond and J. O. Chu, "Extremely high transconductance Ge/Si/sub 0.4/Ge/sub 0.6/ p-MODFET's grown by UHV-CVD," in IEEE Electron Device Letters, vol. 21, no. 3, pp. 110-112, March 2000, doi: 10.1109/55.823572

[44] G. Hellings et al., "High Performance 70-nm Germanium pMOSFETs With Boron LDD Implants," in IEEE Electron Device Letters, vol. 30, no. 1, pp. 88-90, Jan. 2009, doi: 10.1109/LED.2008.2008824

[45] Pillarisetty, R., Chu-Kung, B., Corcoran, S., Dewey, G., Kavalieros, J. P., Kennel, H. W., Kotlyar, R., H, L. V., Lionberger, D., Metz, M. V., Mukherjee, N., Nah, J., Rachmady, W., Radosavljevic, M., Shah, U., Taft, S. R., Then, H. W., Zelick, N. M., & Chau, R. (2010). High mobility strained germanium quantum well field effect transistor as the p-channel device option for low power (Vcc = 0.5 V) III–V CMOS architecture. International Electron Devices Meeting. https://doi.org/10.1109/IEDM.2010.5703312.

[46] Yu, E., Lee, WJ., Jung, J. et al. SiGe Heterojunction FinFET Towards Tera-Hertz Applications. J. Korean Phys. Soc. 72, 527–532 (2018). https://doi.org/10.3938/jkps.72.527

[47] M. Fathipour, B. Abbaszadeh, F. Kohani and F. Farbiz, "The effect of GE mole fraction on the electrical characteristics of nanoscale Si/SiGe heterostructure pMOSFET," 2007 International Semiconductor Device Research Symposium, College Park, MD, USA, 2007, pp. 1-2, doi: 10.1109/ISDRS.2007.4422401

[48] Saha, R., Bhowmick, B. & Baishya, S. Effect of Ge Mole Fraction on Electrical Parameters of Si1−xGex Source Step-FinFET and its Application as an Inverter. Silicon 11, 209–219 (2019). https://doi.org/10.1007/s12633-018-9846-8.

[49] F. Ding, Y. -T. Wu, D. Connelly, W. Zhang and T. -J. K. Liu, "Simulation-Based Study of Si/Si0.9Ge0.1/Si Hetero-Channel FinFET for Enhanced Performance in Low-Power Applications," in IEEE Electron Device Letters, vol. 40, no. 3, pp. 363-366, March 2019, doi: 10.1109/LED.2019.2895323

[50] Atalla, Yousif, Yasir Hashim, & Abdul Nasir Abd. Ghafar. "The impact of channel fin width on electrical characteristics of Si-FinFET." International Journal of Electrical and Computer Engineering (IJECE) [Online], 12.1 (2022): 201-207. Web. 23 Aug. 2024

[51] Saha, R., Bhowmick, B. & Baishya, S. Impact of Mole Fractions due to Work Function Variability (WFV) of Metal Gate on Electrical Parameters in Strained SOI-FinFET. Silicon 12, 577–583 (2020). https://doi.org/10.1007/s12633-019-00163-z

[52] Sahu, S.A., Goswami, R. & Mohapatra, S.K. Characteristic Enhancement of Hetero Dielectric DG TFET Using SiGe Pocket at Source/Channel Interface: Proposal and Investigation. Silicon 12, 513–520 (2020). https://doi.org/10.1007/s12633-019-00159-9

[53] T. Tezuka, S. Nakaharai, Y. Moriyama, N. Sugiyama and S. Takagi, "High-mobility strained SiGe-on-insulator pMOSFETs with Ge-rich surface channels fabricated by local condensation technique," in IEEE Electron Device Letters, vol. 26, no. 4, pp. 243-245, April 2005, doi: 10.1109/LED.2005.844699

[54] F. Ding, Y. -T. Wu, D. Connelly, W. Zhang and T. -J. K. Liu, "Simulation-Based Study of Si/Si0.9Ge0.1/Si Hetero-Channel FinFET for Enhanced Performance in Low-Power Applications," in IEEE Electron Device Letters, vol. 40, no. 3, pp. 363-366, March 2019, doi: 10.1109/LED.2019.2895323

[55] Kuheli Roy Barman and Srimanta Baishya 2022 ECS J. Solid State Sci. Technol. 11 045007DOI 10.1149/2162-8777/ac627d

[56] Kumar, K. S., Yadav, P., & Sharma, S. (2023). Impact of Ge Mole Fraction on the Performance of Si1-xGex/InAs Charged Plasma-Based JLTFET. 18–21. https://doi.org/10.1109/DICCT56244.2023.10110175

CHAPTER
7

Radiation Performance of SOI VSTB FETs under Heavy Ion Radiation

7.1 Introduction

As semiconductor devices continue to be deployed in critical applications such as space technology, avionics, and nuclear systems, their exposure to harsh radiation environments has become a major concern [1-5]. Among the various device structures developed to mitigate short-channel effects and leakage currents, SOI VSTB FETs have shown remarkable potential due to their superior electrostatic control and reduced parasitic effects [6-9]. However, when subjected to radiation—particularly heavy ion strikes—devices are prone to Single Event Effects (SEEs), including transient disturbances and potential permanent failures [10-14]. While SOI inherently provides some radiation hardness by isolating the active region from the substrate, the unique vertical architecture of the VSTB FET introduces additional sensitivity to radiation-induced charge collection. Understanding the radiation response of SOI VSTB FETs is essential for determining their suitability for next-generation radiation-hardened electronics [8, 15-18].

This chapter presents a comprehensive simulation-based investigation of SOI VSTB FETs subjected to heavy ion radiation using TCAD Sentaurus [19]. The focus is on analyzing critical radiation response parameters such as transient drain current, collected charge, and bipolar gain across different ion incidence angles and LET (Linear Energy Transfer) values. The influence of radiation on the device's internal physics, including charge deposition and transport, is carefully examined to understand how radiation interacts with the device geometry and material configuration. The work highlights regions of increased vulnerability within the device, compares responses at various operating conditions, and identifies strategies for mitigating radiation-induced degradation. The insights gained from this study are critical for advancing SOI VSTB FETs in applications, where reliability under radiation exposure is mandatory, such as satellites, defense systems, and nuclear instrumentation.

This chapter is organized as follows: Section 7.2 describes the detailed device structure, simulation methodology, and radiation modeling strategy employed in this work. Section 7.3 presents the results and discussion, focusing on how varying LET values, ion strike locations, and angles of incidence affect the device performance. Section 7.4 concludes the chapter by summarizing key findings and their implications for designing radiation-tolerant SOI VSTB FETs.

7.2 Device Descriptions and Simulation Set-up

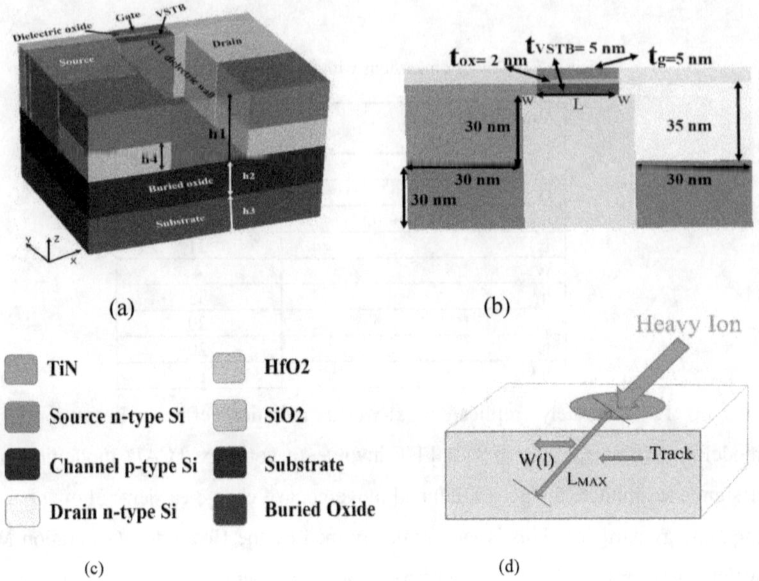

FIGURE 7.1 (a) 3D view of SOI VSTB FET, (b) 2D top view, (c) Regions name, and (d) Device under heavy ion radiation effect.

The SOI VSTB FET utilizes high-performance materials, including HfO_2 with a dielectric constant k=22 and a gate oxide thickness (t_{ox}=2 nm), alongside a TiN gate material. These material choices contribute to superior dielectric behaviour and device stability, pushing the boundaries of current semiconductor technology. Dimensional parameters for the device adhere to the guidelines outlined in the International Technology Roadmap for Semiconductors (ITRS), ensuring industry compatibility and process uniformity [20-21]. Table 7.1 provides a detailed summary of key operational parameters, offering a comprehensive view of the SOI VSTB FET design and functionality. The simulation of radiation effects in SOI VSTB FET leverages a combination of advanced modeling techniques to analyze the performance of device under conditions of high ionizing radiation. For this purpose, 3D Sentaurus TCAD, are employed to assess the impact of SEE on device behaviour [6, 19, 22]. SEE simulations, focus on the transient response of device to high-energy particle strikes, which create a dense cloud of electron-hole pairs capable of causing temporary errors or, in extreme cases, permanent damage to the FET [22].

Table 7.1: Device dimensions with their corresponding parameter.

Parameters	**Dimensions (nm)**
t_{body}	5
t_g	5
t_{ox}	2
L	21
w	4
h_1	35
h_2	20
h_3	20
h4	15

To accurately replicate real-world radiation effects, heavy-ion irradiation modelling within the SOI VSTB FET involves a complex TCAD modelling procedure designed to simulate the generation and movement of charge carriers following a strike by high-energy particles. This is largely determined by the Heavy Ion Generation Model, of which the formation of the electron-hole pairs along a certain track within the semiconductor material is defined. A model describing the localized character of the ion track and the momentary nature of the impact is mathematically defined as a Gaussian distribution in space and time to define the generation rate. The most important parameter in this model is the Linear Energy Transfer (LET) which is a measure of the amount of energy that is lost by the ion in every unit path length that it travels through the device [19]. The ShockleyRead-Hall and auger recombination models are used for charge generation and recombination purpose. The drift-diffusion model serves as the foundation for carrier transport analysis, capturing the behavior of electrons and holes under applied electric fields [7, 9]. Additionally, the density gradient model is incorporated to refine calculations of carrier energy distribution within the transport equations. High doping concentrations in the source and drain regions necessitate the use of Fermi-Dirac statistics to capture carrier distribution nuances accurately [7]. For a realistic representation of mobility under radiation-induced stress, the Lombardi mobility model is applied, accounting for high-k dielectric degradation, which is crucial given the use of HfO_2 as the gate oxide [7-8]. Furthermore, the Masetti model is enabled to represent doping. Dependent mobility variations, while velocity saturation and high-field saturation models are activated to capture the carrier velocity effects under intense electric fields generated by SEE events [7, 19]. Given the susceptibility of the VSTB structure to mechanical stress, stress-induced mobility models, such as stress-dependent electron mobility and saturation velocity models, are included to account for stress-induced behaviour from dielectric wall interactions [7].

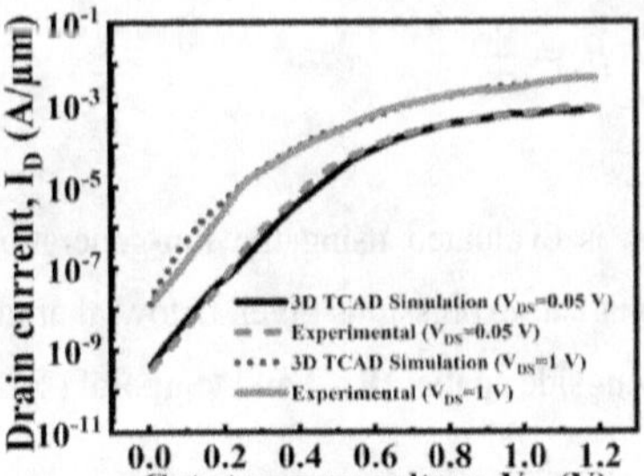

FIGURE 7.2: Calibration of TCAD model against experimental data [54].

Finally, to assess the strain effects induced by structural components and radiation interactions, the deformation potential (DF) model is activated [7]. This model addresses strain-induced alterations in the conduction and valence band edges, providing insights into how these factors impact device resilience. By combining these models, the simulation framework provides a comprehensive view of SOI VSTB FET [8] performance under radiation, highlighting key design considerations for improving device resilience in high-radiation environments. A rigorous calibration of the TCAD models against the experimental results was meticulously carried out to obtain realistic and reliable data from the simulator as referenced in [23]. The harmonious calibration of the experimental data with the simulator outputs is eloquently showcased in Fig. 7.2. During simulations, the gate voltage was set to zero, representing the OFF state of the device. The single-event effects (SEE) is more pronounced at OFF state than ON state, as it influences the electric field and leads to excess charge collection in the device. Heavy-ion irradiation in the SOI VSTB FET is modeled using a carrier generation function that follows a Gaussian distribution with a 10 nm radius and a characteristic time (T_C) of 10 ps. The Linear Energy Transfer (LET) represents charge deposition per unit length, measured in MeV·cm²/mg. Simulations target various incidence angles in the OFF state (V_{DS} = 0.5 V, V_{GS} = 0 V because excess charge collection makes Single Event Effects (SEE) more pronounced than in the ON state. The collected charge (Q_C) is determined by integrating the drain current over the transient time, representing the total charge generated and sensed by the device. Finally, the bipolar gain (β) is calculated as the ratio of Q_C to the deposited charge and deposited charge (Q_D) expressed mathematically as:

$$Q_c = \int_0^t I_d \, dt, \quad \beta = \frac{Q_c}{Q_d} \tag{1}$$

The impact of heavy-ions is evaluated using the ion-generation rate (G_{rion}), a parameter calculated through mathematical expressions given below, that incorporate the linear energy transfer density (G_{rLET}) alongside spatial (R_{ion}) and temporal (Tion) distribution functions

$$G_{rion} = G_{rLET} R_{iON} T_{iON} \tag{2}$$

Here, G_{rLET} is the linear energy transfer density, R_{ion} and T_{ion} are functions for spatial and temporal distributions of the generation rate, respectively,

$$G_{rLET} = \frac{1}{\pi {r_0}^2} LET\_f(l) \tag{3}$$

$$R_{iON(r,l)} = exp\left(-\left(\frac{r}{r_0}\right)^2\right) \tag{4}$$

$$T_{ion}(t) = \frac{2\exp\left(-\left(\frac{t-t_0}{\sqrt{2}W_0}\right)^2\right)}{\sqrt{2}W_0\sqrt{\pi}\left(1+erf\left(\frac{t_0}{\sqrt{2}W_0}\right)\right)} \tag{5}$$

LET_f is defined as heavy-ion model in PicoCoulomb. The r is the perpendicular radius distance from the ion track and t is the time. Additionally, r_o and w_o are the characteristic parameters of Gaussian Function [22]

7.3 Results and Discussion

7.3.1 Ion Strike different Position

The analysis of the electrostatic potential before and after ion strike (LET = 20 MeV-cm^2/mg) reveals critical insights into the radiation sensitivity of the device. The potential curve depicted in Fig. 7.3(b) shows that the increase in the potential barrier after an ion strike in an SOI VSTB FET is primarily due to charge accumulation, floating body effects, and electrostatic field redistribution. When a heavy ion strikes the device,

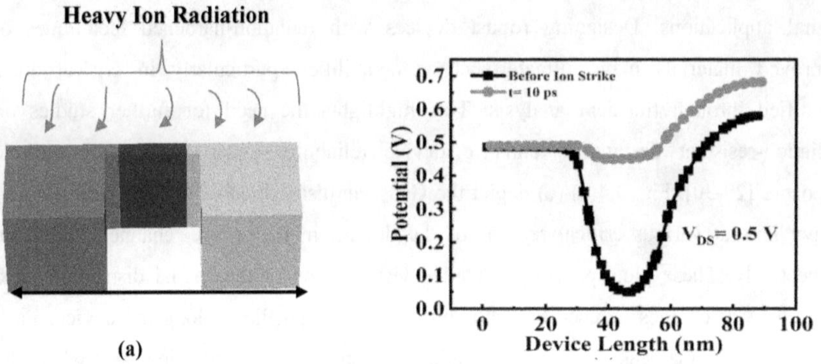

FIGURE 7.3 (a) Heavy-Ion Radiation at different position in X direction and (b) Electrostatic Potential Curve before and after ion strike.

it generates electron-hole pairs in the body, but due to the absence of a bulk substrate in SOI technology, these charges cannot easily dissipate. The generated holes accumulate in the thin body region, leading to a local increase in potential, a phenomenon known as the floating body effect [24-25]. The super-thin body enhances gate control, making the device more sensitive to charge accumulation, which further shifts the electrostatic potential and increases the effective potential barrier. The BOX layer prevents charge leakage into the substrate, forcing excess holes to remain trapped in the body, which further raises the local body potential and increases the potential barrier [26]. Additionally, the accumulation of charge modifies the band structure, causing band bending that increases the energy barrier for electron conduction. At t=10 ps, the potential remains elevated due to slow charge recombination, and while diffusion gradually restores the original potential profile, continuous radiation exposure can lead to long-term reliability issues [27]. This behaviour is critical in radiation-hardened electronics, as charge build up can degrade device performance, emphasizing the need for advanced design strategies to mitigate radiation effects in SOI-based devices.

The sharp potential rise in the radiation-sensitive region between 34 and 55 nm suggest, where the device is most vulnerable to charge perturbations [28]. The additional transient analysis across various device positions helps to identify this critical region, which aligns with the area of maximum ionization due to the radiation strike. The findings imply that the radiation-induced charge generation plays a pivotal role in altering the electrostatic landscape of the device, making it susceptible to unwanted conduction or soft errors in digital applications. Designing robust devices with radiation-hardened techniques or improved materials may mitigate such vulnerabilities, particularly in critical areas identified through transient analysis. This highlights the need for further studies on radiation-resistant designs to enhance device reliability under high-energy particle exposure [29-30]. Fig. 7.4(a)-(c) depict the Gaussian distribution of charges generated by heavy ions at various critical regions of the device: in the source, channel, and drain respectively. These plots provide a detailed visualization of the spatial distribution and magnitude of charges induced by radiation at different positions along the device. Each plot shows that the peak of charge generation corresponds to the specific region where the ion strike occurs, indicating a localized response to the radiation. Fig. 7.4(d) consolidates the data to highlight the greatest charge generation observed at each relevant location.

The analysis reveals a consistent trend: as the point of radiation moves progressively from the source terminal to the drain terminal, the peak of the Gaussian distribution shifts correspondingly along the device length. This demonstrates that the charge generation is highly position-dependent, with the intensity and location of the peak directly tied to the region impacted by the ion strike. The middle of the channel is most vulnerable region to radiation.

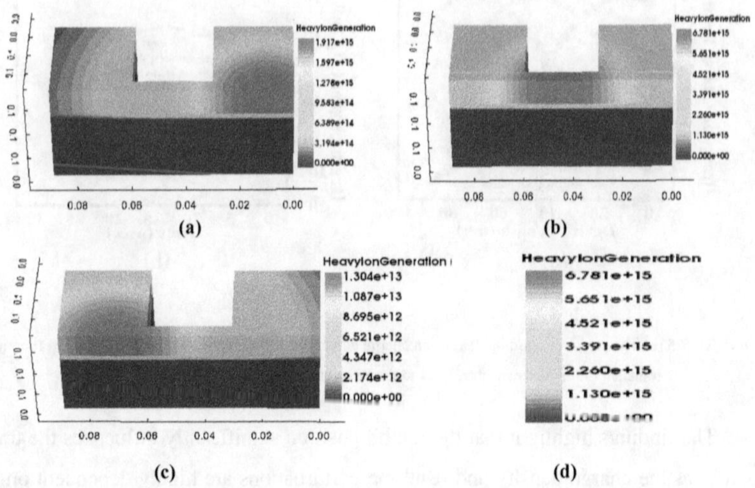

FIGURE 7.4 (a) Source (X = 10 nm), (b) Channel (X = 45 nm), (c) Dain (X = 75 nm), and (d) Heavy Ion-Generation at Constant LET = 20 MeVcm2/mg.

This behaviour emphasizes the localized nature of the radiation-induced effects, which are most pronounced at the precise site of ion impact. Such findings are crucial for understanding the vulnerability of different regions of the device to radiation [31]. They underscore the importance of incorporating design strategies that mitigate these localized effects, such as shielding critical areas or using materials with better radiation tolerance, to ensure device reliability in radiation-prone environments. Fig. 7.5(a) illustrates the variation in heavy-ion charge density as a function of the radiation position across the device. The analysis reveals that the charge density reaches its peak precisely at the point of radiation impact, signifying the localized nature of charge generation induced by heavy-ion strikes. This peak indicates the maximum ionization and charge carrier creation occurring directly at the radiation site, with the charge density decreasing as one moves away from this point [32-33]. The study further examines the transient behavior of the device for different ion impact positions while maintaining a constant LET of 20 MeV-cm^2/mg. Transient analysis was conducted at OFF state of the device, where a drain voltage of 0.5 V was applied to observe how the device responds to ion-induced disturbances.

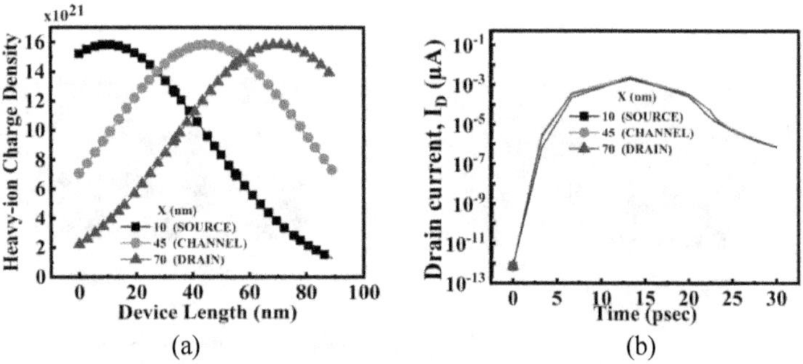

FIGURE 7.5 (a) The peak variation in Heavy-ion charge density at different positions and (b) The variation in the transient drain current with respect to the Position (x=45 nm) strikes at different LET.

The findings highlight that the ion hit position significantly influences the transient response, as the charge density and resulting perturbations are highly dependent on where the radiation occurs within the device [34-35]. This analysis underscores the importance of understanding the spatial effects of radiation on charge dynamics in semiconductor devices. It provides insights into the sensitivity of specific regions, particularly under low-power operating conditions, and highlights the necessity for radiation-hardened design strategies to ensure device robustness against transient disruptions caused by heavy-ion strikes [35]. When an ion strikes the device, it induces a transient current that increases sharply over a period of approximately 12 ps, reaching its peak at 15 ps. This transient response reflects the rapid generation of excess charge carriers caused by ionization. Fig. 7.5(b) illustrates how the drain current varies depending on the position of the ion strike along the device. Notably, the drain current is at its maximum when the radiation occurs at x=45 nm, corresponding to the centre of the channel region [36-37]. Conversely, the drain current is at its minimum for x=10 nm, near the source terminal. The maximum drain current observed at the channel centre is attributed to the generation of a large number of electron-hole (e-h) pairs due to the high ionization rate in this region. The central location of the channel facilitates efficient charge collection and transport, resulting in a pronounced current peak [37]. In contrast, when the radiation occurs closer to the source, the ionization effects are less impactful on the overall device current, leading to a lower drain current. These findings emphasize the critical role of the channel region in determining the device's transient response to radiation [37-38]. The sensitivity of the channel to ion strikes highlights the need for robust design considerations, particularly in applications where radiation exposure is expected. Understanding the spatial dynamics of charge generation and current response can aid in developing strategies to mitigate radiation-induced disruptions and enhance device reliability.

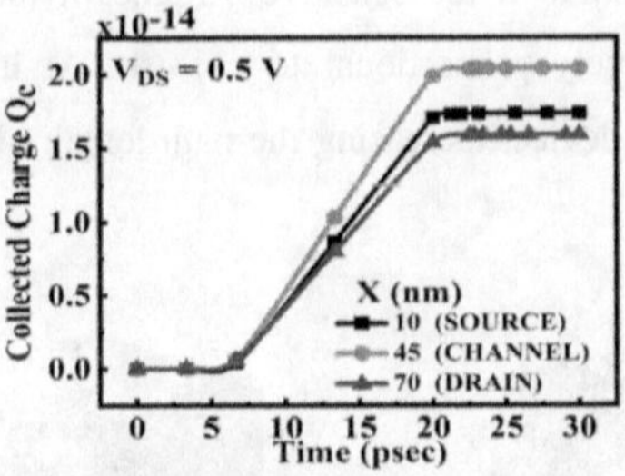

FIGURE 7.6 The collected charge (pC/μm) for different positions at constant LET = 20 MeV-cm^2/mg

Fig. 7.6 illustrates the collected charge (Q_C) over time at three distinct positions within the device: the source (x=10 nm), channel (x=45 nm), and drain (x=70 nm), under a drain-source voltage (V_{DS} = 0.5 V). The data reveals that the source region and drain region (x=10 nm, 70 nm) accumulates the least charge due to the relatively low charge carrier density and weaker electric field in this area. In contrast, the highest collected charge is observed at the centre of the channel region (x=45 nm), where the strongest electric field and highest charge carrier density exist. This aligns with device physics, as the channel region experiences maximum radiation-induced charge collection due to the presence of high-density states and the enhanced movement of ionic particles over longer distances [39-40]. At the drain (x=70 nm), the collected charge is slightly less than at the channel but still substantial, indicating significant radiation interaction. The integration of transient drain current over time quantifies the collected charge, further demonstrating that the channel region is the most radiation-sensitive due to the combined effects of electric field strength, carrier density, and particle travel distance. This analysis underscores the vulnerability of the channel region, particularly at x=45 nm, to radiation-induced damage [40-41]

7.3.2 I_{on} Strike at different Angle of Incidence

The analysis of ion strikes at varying angles of incidence reveals that the angle significantly impacts the ion generation rate and the resulting transient response of the device. Fig. 7.7(a)-(d) compare the ion generation rates for incidence angles of 90^0, 45^0, 30^0, and 0^0, illustrating that the ion generation rate decreases as the angle of incidence reduces from 90^0 to 0^0 (parallel incidence). This behaviour is attributed to the relationship between the charge deposited in the sensitive volume of the device and the angle of incidence, which is inversely proportional to cos (θ). At larger angles, the radiation penetrates deeper into the device, increasing the path length of interaction and enhancing the ion generation rate.

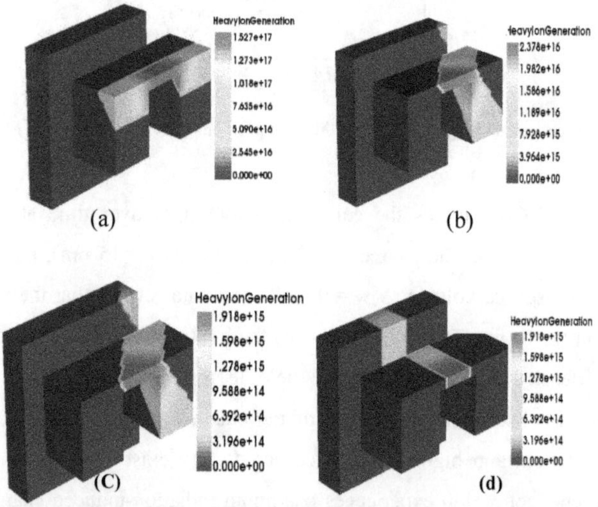

FIGURE 7.7 Heavy Ion-Generation Contour plot with respect to the different angle of incidence (a) 0^0, (b) 30^0, (c) 45^0, and (d) 90^0 at LET = 20 MeV-cm^2/mg.

As the angle of incidence increases, the number of charge carriers generated also rises, leading to higher transient currents and greater collected charges [42-43]. This is evident in Fig. 7.8(a) and Fig. 7.8(b), which show the corresponding increases in collected charge and drain current, respectively, as the angle of incidence becomes steeper. The longer path length of radiation at higher angles enhances ionization, resulting in a higher probability of charge generation within the sensitive volume of the device. Additionally, the angle of incidence varies depending on the type of radiation, with heavier ions being more sensitive to angular effects compared to lighter particles, such as alpha particles. This indicates that the transient response and charge collection are more pronounced under steep-angle ion strikes, making normal incidence (90^0) the most impactful configuration. The analysis highlights the angular sensitivity of radiation effects in TFETs, emphasizing the need for angular-specific design considerations to mitigate radiation-induced vulnerabilities in such devices.

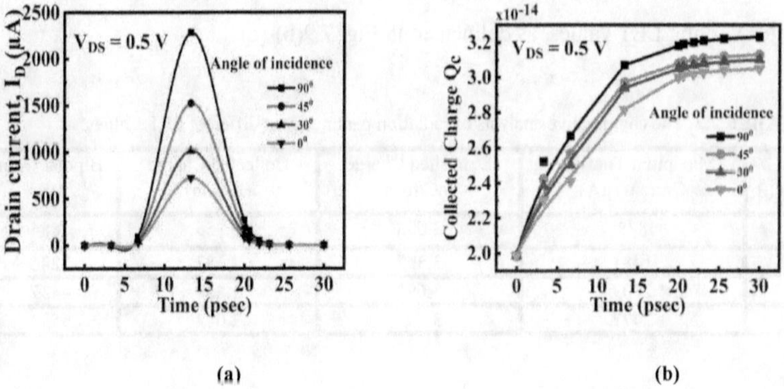

FIGURE 7.8 (a) Transient drain and (b) Collected Charge with respect to angle of incidence at 90^0,45^0, 30^0 and 0^0 at constant LET = 20 MeV-cm^2/mg.

7.3.3 Sensitivity at different LET Values

Fig. 7.9(a) illustrates the Gaussian distribution of heavy-ion generation within the device for various linear energy transfer (LET) values, ranging from 5 to 20 MeV-cm^2/mg. Fig. 8.9(a) focuses on the device's most radiation-sensitive position (x=45 nm), showing that the highest heavy-ion charge density, approximately 1.6×10^{21} cm^{-3}, occurs at LET = 20 MeV-cm^2/mg. This trend is attributed to the direct relationship between LET and the rate of ion generation: as LET increases, the energy deposited per unit length grows, leading to a higher charge density within the material. Higher LET values result in more confined energy deposition due to the shorter range of the radiation, concentrating the energy within a smaller volume. This confinement enhances the production of electron-hole pairs and increases the charge density along the radiation path. Furthermore, higher LET ions generate more ionizations as they interact with the material and experience greater nuclear stopping power. This increased stopping power results in higher energy loss due to interactions with the atomic nuclei, further amplifying the accumulation of charge density. These findings underscore the impact of LET on radiation-induced effects in the device, highlighting that regions with higher LET values experience significantly greater ionization and charge carrier generation. Consequently, devices exposed to high-LET radiation require careful design considerations to mitigate the risks of increased charge density and associated transient disruptions, particularly in sensitive regions like the channel centre at x=45 nm. Further investigation elucidates the transient drain current of the device for varying LET values, as delineated in Fig. 7.9(b)-(c).

TABLE 7.2. The comparative analysis of radiation parameters at different LET Values.

LET (MeVcm2/mg)	Maximum Transient Current (μA)	Deposited Charge (aC/μm)	Collected Charge (fC/μm)	Bipolar Gain (ß)
5	878	3.638	6.855	1884
10	1481	3.597	1.182	3286
15	2009	3.693	1.639	4437
20	2477	3.691	2.042	5530

The analysis indicates that the drain current attains values of 2476.66 μA, 2008.76 μA, 1481.02 μA, and 877.73 μA at LET values of 20, 15, 10, and 5, respectively. The modulation of transient current is observed as a consequence of ion impacts, with elevated LET values demonstrating a more pronounced transient current. Moreover, the charge accumulation within the device is directly correlated with the quantity of linear energy transfer imparted to the device following ion strikes [44-48].

TABLE 7.3. Comparison Table of different FET under Radiation effects

S No.	Device	I_{DMax} (mA)	Max Q_C (fC)
1	LTFET[49]	260	2.1
2	DG-FeTFET [50]	272	2.1
3	Conventional TFET [51]	1600	52
4	Doping less TFET [52]	453	2.92
5	Ge-Source TFET [22]	482	34.6
This work	SOI VSTB FET	1481.02 μA	11.82

Consequently, radiation exposure at higher LET values results in an increased amount of collected charge, as illustrated in Fig. 7.9(d). Table 7.2 delineates the comparative analysis of radiation parameters across different LET values. It is observed that as the LET increases, there is a concomitant elevation in the values of transient current, collected charge, and deposited charge. This study examines the heavy ion radiation effects on the SOI VSTB FET at V_{DS}=0.5 V, as shown in Table 7.3 for LET = 10 MeV-cm²/mg. The SOI VSTB FET in this work demonstrates drain current (I_{Dmax}=1481.02 μA) and maximum collected charge (Q_{Cmax}=11.82 fC) compared to conventional TFETs, indicating improved radiation resilience. The device operates at a drain voltage of V_{DS}=0.5 V, aligning with most TFETs except the conventional TFET, which requires 1.5 V. The reduced collected charge suggests fewer electron-hole pairs are generated, thereby mitigating the single-event transient (SET) effect.

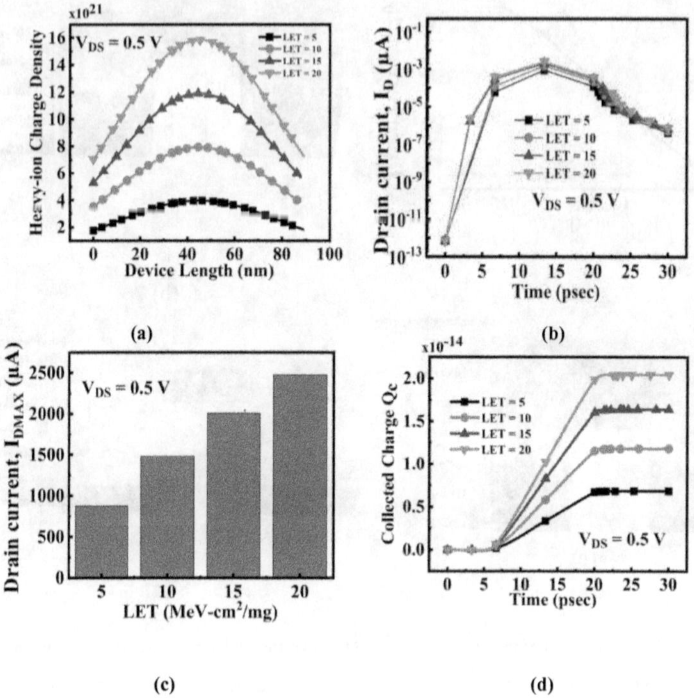

FIGURE 7.9 (a) The variation in Heavy Charge density with respect to position X = 45 nm at different LETs and (b) The Variation of transient Drain current shown for different LETs with respect to time, (c) Maximum transient Drain current, and (d) Collected charge(Q_C) with respect to different LET Value.

Although double-gate TFETs exhibit higher susceptibility to radiation due to their larger exposed area and increased gate oxide-semiconductor interfaces, leading to charge trapping and radiation-induced effects, the SOI VSTB FET structure enhances immunity against such issues. With a better balance between performance and robustness, this device provides a competitive edge by reducing sensitivity to radiation, while maintaining high current efficiency, making it a promising candidate for radiation-hardened applications.

7.3.4 Alpha Particle Generation

The comparative analysis of the effects of heavy-ion radiation and alpha particles on SOI

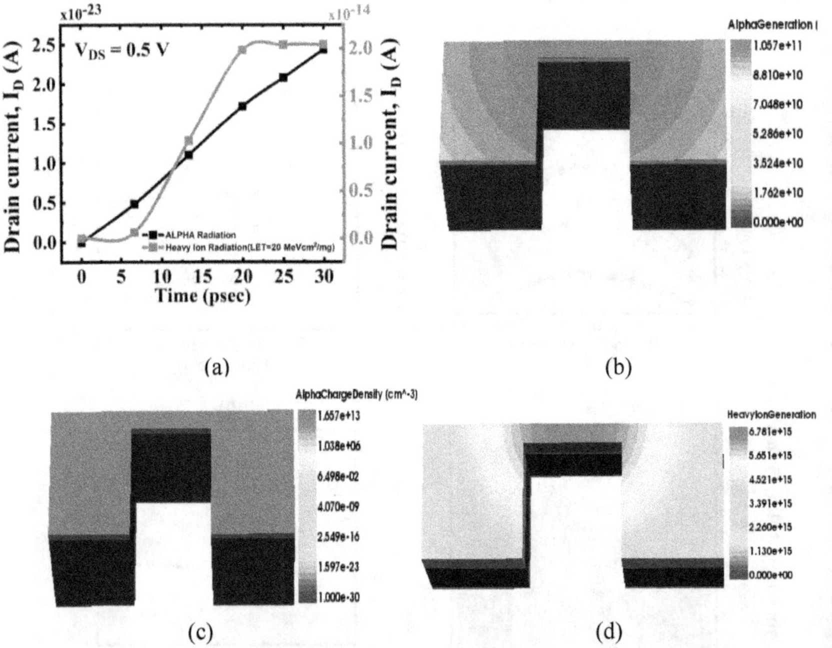

FIGURE 7.10 (a) The Variation of transient Drain current shown for Alpha particle and heavy ion radiation with respect to time (b). Alpha Particle Generation (c) Alpha charge Generation and (d) Heavy ion generation at Constant LET = 20 MeV-cm2/mg

VSTB FET devices, as demonstrated in the plot (Fig. 7.10(a)), reveals that heavy ions are significantly more effective in causing damage to the device than alpha particles. The graph shows that the drain current (I_D) under heavy-ion radiation (LET = 20 MeV·cm²/mg) Increases more sharply compared to alpha particle exposure, with approximately 10^4 times more charge deposition observed for heavy ions. This higher ionization density occurs due to the greater charge deposition along the radiation trajectory of heavy ions, which induces a more pronounced increase in drain current over time. For instance, while both radiation types exhibit a rising trend in drain current (I_D) within 30 ps, the current associated with heavy ions reaches saturation at a significantly higher value than that under alpha particles [19, 55]. These observations highlight the superior capability of heavy ions to induce higher ionization and consequently, greater damage to the SOI VSTB FET device. Conversely, alpha particles, while still impactful, cause less ionization and charge deposition, resulting in comparatively reduced degradation in device performance as shown in Fig. 7.10 (b)-(d). This underscores the critical need for designing robust SOI VSTB FETs to withstand high-energy particle radiation environments effectively.

7.4 Conclusion

The sensitivity of SOI VSTB FET to heavy-ion radiation has been investigated across source, channel and drain positions, with results showing maximum sensitivity at the intermediate channel position, where an ion strike shifts the device from the OFF state to the ON state. Further analysis, varying the angle of ion incidence (0^0, 30^0, 45^0, and 90^0), reveals that increasing the angle leads to greater charge accumulation on the device surface. Radiation effects were also studied by varying the LET, with the device at LET = 20 MeV-cm²/mg exhibiting a peak transient DC current of 2476.66 μA and a collected charge of 3.59×10^{-18} C/μm. The bipolar gain exceeding one indicates high sensitivity to ionic radiation, making the device prone to damage and transient errors. Heavy-ion radiation causes significantly greater damage to SOI VSTB FETs than alpha particles, depositing approximately 10^4 times more charge due to higher ionization density along its trajectory. These findings are critical for high-energy applications, such as atomic and nuclear operations, and have significant implications for space exploration, where understanding and mitigating heavy-ion radiation effects are essential for ensuring the reliability of electronic systems and astronaut safety. This research highlights the need for robust device design to withstand radiation in demanding environments.

References

[1] Ma, T. P. & Dressendorfer, P. V. Ionizing Radiation Effects in MOS Devices and Circuits (Wiley, 1989).

[2] Paccagnella, A., Cester, A. & Cellere, G. Ionizing radiation effects on MOSFET thin and ultra-thin gate oxides. In IEDM Technical Digest. IEEE International Electron Devices Meeting, 473–476 (IEEE, 2004).

[3] Arshak, K.; Korostynska, O. Response of Metal Oxide Thin Film Structures to Radiation. Mater.Sci. Eng. B 2006, 133, 1–7. doi:10.1016/j.mseb.2006.06.012

[4] Ma, T.P.; Dressendorfer, P.V. Ionizing Radiation Effects in MOS Devices and Circuits; John Wiley & Sons: New York, 1989.

[5] Schwank, J.R.; Shaneyfelt, M.R.; Dodd, P.E. Radiation Hardness Assurance Testing of Microelectronic Devices and Integrated Circuits: Radiation Environments, Physical Mechanisms, and Foundations for Hardness Assurance. IEEE Trans. Nucl. Sci. Jun. 2013, 60 (3), 2074–2100. doi:10.1109/TNS.2013.2254722.

[6] V. Koldiaev and R. Pirogova, "Vertical super-thin body semiconductor on dielectric wall devices and methods of their fabrication", U.S. Patent 8,796,085 B2, 5 Aug 2014.

[7] S. Roy, A. Chatterjee, D.K. Sinha, R. Pirogova, and S. Baishya, "2-D analytical modeling of surface potential and threshold voltage for vertical super-thin body FET", IEEE Trans. Electron Devices, vol. 64, no. 5, pp. 2106–2112, 2017,

[8] Roy Barman, Kuheli & Baishya, Srimanta. (2022). Silicon and Germanium Vertical Super-Thin Body (VSTB) FET: A Comparative Performance Overview Including Architectural Stress-Strain Impact. ECS Journal of Solid State Science and Technology. 11. 10.1149/2162-8777/ac627d

[9] R. Barman and S. Baishya, "Performance analysis of vertical super-thin body (VSTB) FET and its characteristics in presence of noise", Appl. Phys. A, vol. 125, pp. 401, 2019, https://doi.org/10.1007/s00339-019-2682-x

[10] Q. Wang, H. Liu, S. Wang, and S. Chen, ''TCAD simulation of single event-transient effects in L-shaped channel tunneling field-effect transistors,'' IEEE Trans. Nucl. Sci., vol. 65, no. 8, pp. 2250–2259, Aug. 2018, doi: 10.1109/TNS.2018.2851366

[11] Electron Irradiation on Metal Oxide Silicon Field Effect Transistors in Space Applications - Sujatha R - IJFMR Volume 6, Issue 5, September-October 2024. DOI 10.36948/ijfmr.2024.v06i05.28434

[12] Tian J, Cao R, Liu Y, Cai Y, Mei B, Zhao L, Cui S, Lv H, Xue Y. Electron-Induced Single-Event Effect in 28 nm SRAM-Based FPGA. Electronics. 2024; 13(12):2233. https://doi.org/10.3390/electronics13122233

[13] Grome, C.A.; Ji, W. A Brief Review of Single-Event Burnout Failure Mechanisms and Design Tolerances of Silicon Carbide Power MOSFETs. Electronics 2024, 13, 1414. https://doi.org/10.3390/electronics13081414

[14] Yu, Hao, Wei Zhou, Hongxia Liu, Shulong Wang, Shupeng Chen, and Chang Liu. 2024. "Degradation Induced by Total Ionizing Dose and Hot Carrier Injection in SOI FinFET Devices" Micromachines 15, no. 8: 1026. https://doi.org/10.3390/mi15081026

[15] Liu, Baojun, Ping Zhou, and Liang Qian. "Single Event Transient of SOI FinFET with Total Ionizing Dose Irradiation." Vibroengineering Procedia 51 (October 20, 2023): 76–81. https://doi.org/10.21595/vp.2023.23602.

[16] Y. Sun et al., "SOI FinFET Design Optimization for Radiation Hardening and Performance Enhancement," in IEEE Transactions on Device and Materials Reliability, vol. 23, no. 3, pp. 386-394, Sept. 2023, doi: 10.1109/TDMR.2023.3287839.

[17] V. A. Pan et al., "Charge Collection in SOI Microdosimeters and Their Radiation Hardness," in IEEE Transactions on Nuclear Science, vol. 70, no. 4, pp. 568-574, April 2023, doi: 10.1109/TNS.2023.3242267.

[18] Y. Huang et al., "Total Ionizing Dose Radiation Effects Hardening Using Back-Gate Bias in Double-SOI Structure," in IEEE Transactions on Nuclear Science, vol. 69, no. 3, pp. 453-461, March 2022, doi: 10.1109/TNS.2022.3145027.

[19] TCAD Sentaurus Device User Manual, Version J 2022.12 (Synopsis Inc.)

[20] F. Balestra, Silicon-on-Insulator Devices, Wiley Encyclopedia of Electrical andElectronics Engineering,edited by John G. Webster (Wiley, 2014).

[21] International Technology Roadmap for Semiconductors," 2020. http://www.itrs.net

[22] S. Tiwari, R. Saha, G. Gopal and T. Varma, "Sensitivity Analysis of Ge Source Double Gate TFET Under Heavy Ion Irradiation," in IEEE Sensors Journal, vol. 24, no. 22, pp. 36730-36736, 15 Nov.15, 2024, doi: 10.1109/JSEN.2024.3463870.

[23] Y. Morita, R. Tsuchiya, T. Ishigaki, N. Sugii, T. Iwamatsu, T. Ipposhi, H. Oda, Y. Inoue, K. Torii, and S. Kimura, “Smallest Vth Variability Achieved by Intrinsic Silicon on Thin BOX (SOTB) CMOS with Single Metal Gate,” in Proc. of Symposium on VLSI Technology (VLSIT), pp. 166 – 167, 2008.
[24] S. B. Chiah and X. Zhou, "Floating-Body Effect in Partially/Dynamically/Fully Depleted DG/SOI MOSFETs Based on Unified Regional Modeling of Surface and Body Potentials," in IEEE Transactions on Electron Devices, vol. 61, no. 2, pp. 333-341, Feb. 2014, doi: 10.1109/TED.2013.2288309.
[25] M. Gerold, G. Dollinger, L. Günther, J. Reindl, M. Rüb,Sensitivity study of super-junction power MOSFETs by spatial and depth resolved heavy ion SEE mapping with various bias, LETs and ion ranges,Microelectronics Reliability,Volume 155,2024,115309,ISSN 0026-2714, https://doi.org/10.1016/j.microrel.2023.115309
[26] C. Peng et al., "Influence of the Total Ionizing Dose Irradiation on 130 nm Floating-body PDSOI NMOSFETs," in IEEE Transactions on Nuclear Science, vol. 62, no. 1, pp. 314-322, Feb. 2015, doi: 10.1109/TNS.2015.2390636.
[27] C. -H. Yu, Y. Wang, M. -T. Bao, X. -J. Li, J. -Q. Yang and F. Cao, "Impact of Heavy-Ion Irradiation in an 80-V Radiation-Hardened Split-Gate Trench Power UMOSFET," in IEEE Transactions on Electron Devices, vol. 69, no. 2, pp. 664-668, Feb. 2022, doi: 10.1109/TED.2021.3135369.
[28] C. Spoorthi, J. Grace, A. Chavan and C. S. M. Rao, "Total Ionizing Dose and Radiation Particle Strike Analysis of Nanoscale FinFET Devices and Circuits," 2020 4th International Conference on Electronics, Communication and Aerospace Technology (ICECA), Coimbatore, India, 2020, pp. 479-484, doi: 10.1109/ICECA49313.2020.9297573.
[29] H. S. Gupta et al., "Design and Analysis of 8T Radiation Hardened SRAM using 65nm Process," 2024 28th International Symposium on VLSI Design and Test (VDAT), Vellore, India, 2024, pp. 1-4, doi: 10.1109/VDAT63601.2024.10705718.
[30] Choi, W.C.; Jo, S.-H. Radiation Hardened Read-Stability and Speed Enhanced SRAM for Space Applications. Appl. Sci. 2024, 14, 9015. https://doi.org/10.3390/app14199015
[31] Kaindl, W., Wachutka, G., Sölkner, G. (2001). Modeling and Simulation of Charge Generation Events Caused by Ion Irradiation in High Voltage Power Devices. In: Tsoukalas, D., Tsamis, C. (eds) Simulation of Semiconductor Processes and Devices 2001. Springer, Vienna. https://doi.org/10.1007/978-3-7091-6244-6_101
[32] Elwailly, Adam & Saltin, Johan & Gadlage, Matthew & Wong, Hiu. (2021). Radiation Hardness Study of LG = 20 nm FinFET and Nanowire SRAM Through TCAD Simulation. IEEE Transactions on Electron Devices. PP. 1-6. 10.1109/TED.2021.3067855.
[33] Dubey, Avashesh & Narang, Rakhi & Saxena, Manoj & Gupta, Mridula. (2020). Investigation of Single Event Transient Effects in Junctionless Accumulation Mode MOSFET. IEEE Transactions on Device and Materials Reliability. PP. 10.1109/TDMR.2020.3014176.
[34] Periyasamy, Rajendiran & Raj, Srinivasan. (2019). Single-event radiation performance analysis of junction and junctionless FET-based low-noise amplifiers. Journal of Computational Electronics. 18. 10.1007/s10825-019-01370-y.
[35] Dubey, Avashesh & Singh, Ajay & Narang, Rakhi & Saxena, Manoj & Gupta, Mridula. (2017). Modeling and Simulation of Junctionless Double Gate Radiation Sensitive FET (RADFET) Dosimeter. IEEE Transactions on Nanotechnology. PP. 1-1. 10.1109/TNANO.2017.2719286.
[36] Gaillardin, M. & Raine, M. & Paillet, P. & Martinez, M. & Marcandella, C. & Girard, Sylvain & Duhamel, O. & Richard, N. & Andrieu, Francois & Barraud, S. & Faynot, O.. (2013). Radiation effects in advanced SOI devices: New insights into Total Ionizing Dose and Single-Event Effects. 1-2. 10.1109/S3S.2013.6716530.
[37] D. McMorrow, A. R. Knudson, J. B. Boos, Doe Park and J. S. Melinger, "Ionization-induced carrier transport in InAlAs/InGaAs high electron mobility transistors," in IEEE Transactions on Nuclear Science, vol. 51, no. 5, pp. 2857-2864, Oct. 2004, doi: 10.1109/TNS.2004.835060.
[38] Paul M Haney et al 2015 Nanotechnology 26 295401DOI 10.1088/0957-4484/26/29/295401
[39] Rohith Sai, K., K. Girija Sravani, K. Srinivasa Rao, B. Balaji, and Vipul Agarwal. "Design and Performance Analysis of High-k Gate All Around Fin-field Effect Transistor." International Journal of Engineering 37, no.3 (2024): 476-483, doi: 10.5829/IJE.2024.37.03C.04
[40] Autran, J., & Munteanu, D. (2023). New insights into diffusion–collection modeling of radiation-induced charge in semiconductor devices. Journal of Applied Physics, 134(17). https://doi.org/10.1063/5.0156698
[41] Schwank, J. R., Ferlet-Cavrois, V., Dodd, P. E., Shaneyfelt, M. R., Vizkelethy, G., Paillet, P., & Flament, O. (2004). Analysis of heavy-ion induced charge collection mechanisms in SOI circuits. Solid-State Electronics, 48(6), 1027–1044. https://doi.org/10.1016/J.SSE.2003.12.038

[42] Yu, J.-T., Chen, S., Chen, J., & Huang, P. (2016). Simulation analysis of heavy-ion-induced charge collection between nanoscale bulk-Si FinFET and conventional planar device. European Conference on Radiation and Its Effects on Components and Systems, 1–4. https://doi.org/10.1109/RADECS.2016.8093173
[43] N. Kamal, M. Panchore and J. Singh, "3-D Simulation of Junction- and Doping-Free Field-Effect Transistor Under Heavy Ion Irradiation," in IEEE Transactions on Device and Materials Reliability, vol. 18, no. 2, pp. 173-179, June 2018, doi: 10.1109/TDMR.2018.2811493
[44] D. Munteanu et al., "Investigation of Quantum Effects in Ultra-Thin Body Single- and Double-Gate Devices Submitted to Heavy Ion Irradiation," in IEEE Transactions on Nuclear Science, vol. 53, no. 6, pp. 3363-3371, Dec. 2006, doi: 10.1109/TNS.2006.886206
[45] M. Murat, A. Akkerman and J. Barak, "Electron and Ion Tracks in Silicon: Spatial and Temporal Evolution," in IEEE Transactions on Nuclear Science, vol. 55, no. 6, pp. 3046-3054, Dec. 2008, doi: 10.1109/TNS.2008.2007646.
[46] Y. Sun, J. Shao, Z. Liu, X. Li, Y. Liu, and Y. Shi, ''Evaluation of single event-transient effects in reconfigurable field effect transistor beyond 3 nm technology node,'' IEEE Trans. Electron Devices, vol. 68, no. 12, pp. 6001–6006, Dec. 2021, doi: 10.1109/TED.2021.3119006
[47] B. Ye, L.-H. Mo, P.-F. Zhai, L. Cai, T. Liu, Y.-N. Yin, Y.-M. Sun, and J. Liu, ''Impact of heavy ion energy and species on single-event upset in commercial floating gate cells,'' Microelectron. Rel., vol. 120, May 2021, Art. no. 114128, doi: 10.1016/j.microrel.2021.114128.
[48] Y. Morita, R. Tsuchiya, T. Ishigaki, N. Sugii, T. Iwamatsu, T. Ipposhi, H. Oda, Y. Inoue, K. Torii, and S. Kimura, "Smallest Vth Variability Achieved by Intrinsic Silicon on Thin BOX (SOTB) CMOS with Single Metal Gate," in Proc. of Symposium on VLSI Technology (VLSIT), pp. 166 – 167, 2008.
[49] H. Liu, M. Cotter, S. Datta, and V. Narayanan, ''Soft-error performance evaluation on emerging low power devices,'' IEEE Trans. Device Mater. Rel., vol. 14, no. 2, pp. 732–741, Jun. 2014, doi: 10.1109/TDMR.2014. 2316505
[50] Colinge, Jean-Pierre & Lee, C & Afzalian, Aryan & Dehdashti Akhavan, Nima & Yan, Ran & Razavi, Pedram & O'Neill, Brendan & Blake, Alan & White, Mary & Kelleher, Anne-Marie & Mccarthy, Brendan & Murphy, Richard, "Nanowire transistors without junctions", Nature nanotechnology, 2010, 5. 225-9. 10.1038/nnano.2010.15.
[51] Q. Wang, H. Liu, S. Wang, and S. Chen, ''TCAD simulation of singleevent-transient effects in L-shaped channel tunneling field-effect transistors,'' IEEE Trans. Nucl. Sci., vol. 65, no. 8, pp. 2250–2259, Aug. 2018, doi: 10.1109/TNS.2018.2851366.
[52] G. Tian, J. Bi, G. Xu, K. Xi, X. Yang, H. Yin, Q. Xu, and W. Wang, ''Heavy ion induced single-event-transient effects in nanoscale ferroelectric vertical tunneling transistors by TCAD simulation,''Semicond. Sci. Technol., vol. 35, no. 10, Oct. 2020, Art. no. 105010, doi: 10.1088/1361- 6641/aba549.
[53] Y. Wu and Y. Takahash, '' The impact of tunnel FET on heavy ion induced transient effect,'' in Proc. ECS Meeting. Chesapeake, VA, USA, Oct. 2017, doi: 10.1149/MA2017-02/26/1143
[54] M. Sharma, R. Narang, M. Saxena, and M. Gupta, ''Single event transient effect on tapered angle hetero-junction dopingless TFET for radiation sensitive applications,'' in Proc. IEEE VLSI Device Circuit Syst. (VLSI DCS), Feb. 2022, pp. 261–264, doi: 10.1109/VLSIDCS53788.2022.9811459.
[55] Pankaj Kumar, Kalyan Koley, and Subindu Kumar. "Comparative analysis of heavy ions and alpha particles impact on gate-all-around TFETs and gate-all-around MOSFETs." Micro and Nanostructures, vol. 192, pp. 207875, 2024, doi.org/10.1016/j.micrna.2024.207875

CHAPTER

8

Radiation Performance of 14 nm SiC-4H SOI VSTB FETs under Heavy Ion Radiation

8.1 Introduction

As semiconductor devices are increasingly deployed in harsh environments like outer space, nuclear reactors, and avionics systems, understanding their behavior under radiation has become critically important. Radiation exposure, particularly from heavy ions, can severely impact the performance and reliability of semiconductor devices by causing Single Event Effects (SEEs), such as transient faults and permanent failures [1-5]. While Silicon-On-Insulator (SOI) technology inherently provides better isolation and reduced leakage compared to bulk silicon, combining it with 4H-Silicon Carbide (4H-SiC) material offers further advantages, including higher breakdown voltage, superior thermal stability, and improved radiation hardness [6-8]. However, despite these inherent benefits, scaling down to advanced nodes like 14 nm makes these devices more susceptible to radiation-induced charge collection. Therefore, this chapter focuses on evaluating the radiation performance of 14 nm 4H-SiC SOI VSTB FETs under heavy ion radiation using advanced TCAD simulations [9]. The aim is to understand how variations in ion energy (LET), angles of incidence, and material properties influence critical parameters like transient current, collected charge, and bipolar gain.

Inside this chapter, detailed TCAD simulation-based studies are presented to analyze the device response under varying LET conditions. The simulations helps to identify which physical phenomena dominate under radiation exposure and how the unique combination of SOI and 4H-SiC architecture affects device robustness. The chapter also compares performance at different strike positions (source, channel, drain) and explores the effects of ion incidence angles on collected charge and transient current. This comprehensive analysis serves as a guide for optimizing SOI VSTB FETs for radiation-hardened applications, especially in the context of next-generation aerospace and nuclear electronics.

This chapter is structured as follows: Section 8.2 describes the device architecture and simulation methodology for heavy ion radiation on the SOI VSTB FET. Section 8.3 discusses the results, including heavy ion charge density, collected charge (Q_C) for different LET values. Finally, Section 8.4 presents the conclusion of this study.

8.2 Device Descriptions and Simulation Set-up

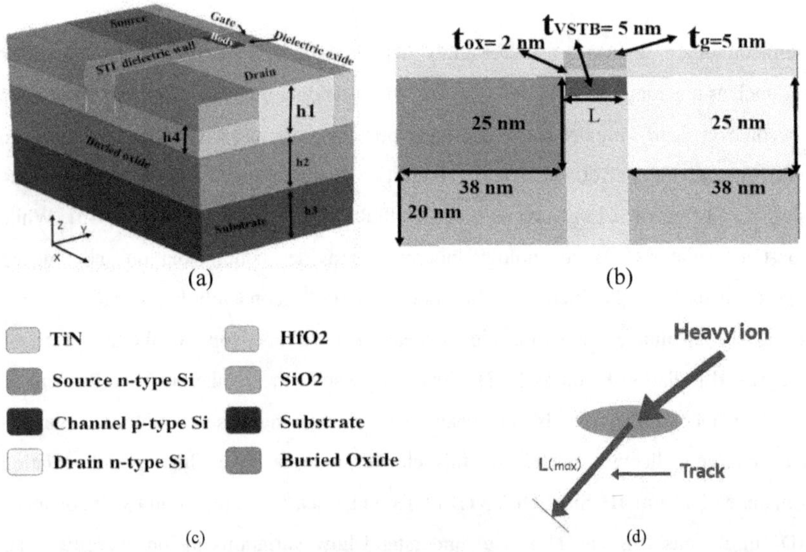

FIGURE 8.1 (a) 3D view of 4H-SiC SOI VSTB FET, (b) 2D top view, (c) Regions name, and (d) Device under heavy ion radiation effect.

Fig. 8.1(a) illustrates 3D schematic of the 4H-SiC SOI VSTB FET, highlighting the advanced structural and design features of the device. Figs. 8.1(b) and 8.1(c) provide 2D top views and color-coded diagrams, respectively, offering detailed insights into the device layout and configuration essential for understanding its spatial and functional characteristics. The fabrication process follows established methodologies reported in prior studies [6, 9-11]. This emphasizes the critical importance of precision and consistency in fabrication to achieve optimal device performance and reliability.

Table 8.1: Device dimensions with their corresponding parameter.

Parameters	Dimensions (nm)
t_{body}	5
t_g	5
t_{ox}	2
L	14
h_1	20
h_2	20
h_3	20
h4	10

The process involves carefully controlled doping concentrations of 10^{19} cm^{-3}, 10^{15} cm^{-3}, 10^{17} cm^{-3}, and 10^{15} cm^{-3} for the source, channel, drain, and substrate regions, respectively [7, 12]. The slightly lower doping concentration in the drain compared to the source is a deliberate design choice aimed at minimizing leakage current (I_{OFF}) and enhancing overall device efficiency [12]. The SOI VSTB FET employs advanced materials, including, SiC-4H as an active regions of the device, HfO_2 with a dielectric constant k=22 and a gate oxide thickness of t_{ox}=2 nm, in combination with a TiN gate electrode. These material choices enable enhanced dielectric performance and improved device stability, contributing to the advancement of semiconductor technology. The dimensional parameters of the device are aligned with the International Technology Roadmap for Semiconductors (ITRS) specifications, ensuring compatibility with industry standards and consistency in manufacturing processes [9, 13].

Table 8.1 presents a detailed overview of the key operational parameters, providing a comprehensive understanding of the design and functionality of the SOI VSTB FET. The simulation of radiation effects in SOI VSTB FET utilizes advanced modelling approaches to evaluate device performance under high ionizing radiation conditions. 3D Sentaurus TCAD are employed to analyze the influence of Single Event Effects (SEE) on device behaviour [6, 14-15]. SEE simulations focus on the transient response of the device to high-energy particle strikes, which generate a dense cloud of electron-hole pairs that can induce temporary malfunctions or, in severe cases, result in permanent damage to the SOI

VSTB FET [15]. To accurately replicate real-world radiation effects, a comprehensive set of physics-based models is activated within the TCAD simulation framework [14-15]. Carrier transport is primarily modelled using the drift-diffusion approach, which effectively captures electron and hole behaviour under applied electric fields [7, 12]. The density gradient model is also incorporated to enhance the accuracy of carrier energy distribution within the transport equations. Due to high doping concentrations in the source and drain regions, Fermi-Dirac statistics are employed to represent carrier distribution accurately [7]. To model mobility under radiation-induced stress conditions, the Lombardi mobility model is utilized, accounting for high-k dielectric degradation relevant to the use of HfO_2 as the gate oxide [7, 12]. Additionally, the Masetti model is enabled to simulate doping-dependent mobility variations, while velocity saturation and high-field saturation models are activated to capture carrier velocity effects under the intense electric fields induced by SEE [7, 15]. Given the mechanical stress sensitivity of the VSTB architecture, stress-dependent models for electron mobility and saturation velocity are included to account for stress effects originating from dielectric wall interactions [7].

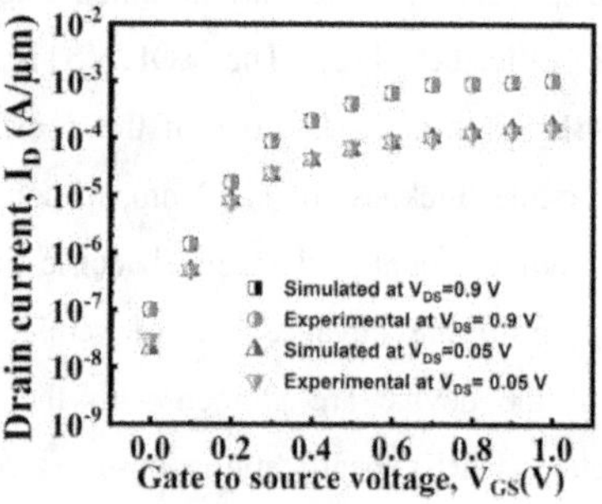

FIGURE 8.2 Calibration of TCAD model against experimental data [42].

To evaluate the strain effects induced by structural components and radiation interactions, the deformation potential (DF) model is enabled [7]. This model accounts for strain-induced modifications in the conduction and valence band edges, offering critical insights into the impact of mechanical stress on device resilience. The integration of all physics-based models within the simulation framework enables a comprehensive assessment of 4H-SiC SOI VSTB FET performance under heavy ion and alpha radiation exposure [16], emphasizing essential design strategies for enhancing radiation tolerance in extreme environments. A meticulous calibration of the TCAD simulation results against experimental data was performed to ensure accuracy and reliability, as detailed in [17]. The consistency between the simulated and experimental outcomes is clearly demonstrated in Fig. 8.2.

8.3 Results and Discussion

8.3.1 Comparative analysis of heavy ion radiation for Si, SiC, and SiC-4H

The analysis of the electrostatic potential after ion strike (LET = 20 MeV-cm²/mg) reveals critical insights into the radiation sensitivity of the device.

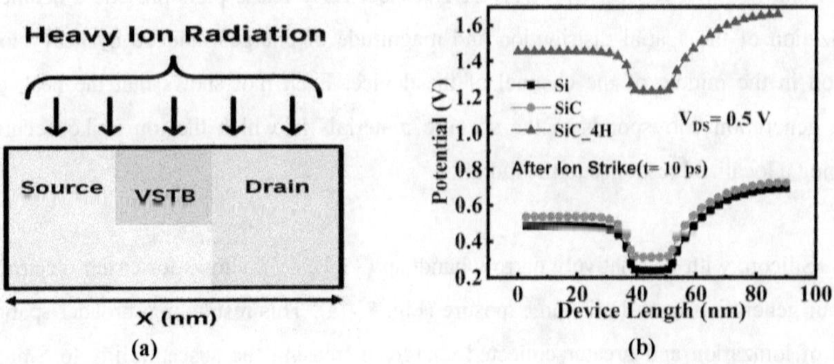

FIGURE 8.3 (a) Heavy-Ion Radiation at different position in X direction and (b) Electrostatic Potential Curve before and after ion strike.

The potential curve is depicted in Fig. 8.3(b) shows that the increase in potential barrier after an ion strike in SOI VSTB FET with SiC-4H compared to SiC and Silicon, which is primarily due to inherent material properties of 4H-SiC, particularly its wide bandgap of approximately 3.26 eV, which significantly suppresses intrinsic carrier generation under high-energy particle irradiation [18-19].

The high critical electric field and lower dielectric constant of 4H-SiC facilitate rapid recombination of radiation-induced electron-hole pairs and restrict charge diffusion, thereby minimizing transient potential collapse. In addition, the superior thermal conductivity of 4H-SiC aids in dissipating localized thermal spikes generated during ion strikes, further contributing to potential stability. In contrast, SiC and Si, with narrower bandgaps and higher carrier motilities, exhibit greater susceptibility to transient charge buildup and diffusion, resulting in a more pronounced drop in potential at the site of impact [19]. Consequently, the elevated and stable potential barrier observed in the 4H-SiC-based device underscores its superior electrostatic control and enhanced immunity to radiation-induced perturbations, highlighting its suitability for high-radiation and space-based applications [19-20]. The sharp potential fall in the radiation-sensitive region between 38 and 52 nm suggests where device is most vulnerable to charge perturbations [21-22]. The additional transient analysis across various device positions helps identify this critical region, which aligns with the area of maximum ionization due to the radiation strike [23-24]. Fig. 8.4(a)-(c) depict the Gaussian distribution of charges generated by heavy ions for Si, SiC, and SiC-4H, based SOI VSTB FET respectively. These plots provide a detailed visualization of the spatial distribution and magnitude of charges induced by heavy ion radiation in the middle of the channel of the device. Each plot shows that the peak of charge generation corresponds to the specific materials in which the ion strike occurs, indicating a localized response to the radiation.

Silicon, with its relatively narrow bandgap (~1.12 eV), allows for easier electron-hole pair generation upon radiation exposure (Fig. 8.4(a). This results in a broader spatial spread of ionization and greater collected charge, increasing the susceptibility to Single Event Effects (SEEs). According to such behaviour leads to significant transient currents, which in advanced CMOS can propagate through digital logic as spurious signals. The literature also affirms that silicon's high carrier mobility supports larger bipolar gain and longer charge collection times, both of which elevate the risk of latch-up and other soft errors

Standard SiC (Fig. 8.4(b)), while better than Si, does not match the superior radiation hardness of the 4H polytype. The slightly narrower bandgap and possible material imperfections can increase carrier generation slightly more than in 4H-SiC. As detailed by Aishwarya & Lakshmi (2024), materials with moderate tunneling barriers and dopant profiles may still produce higher I_{peak} values under intense ion strikes, especially when the ion strike aligns with sensitive interfaces like the drain/channel junction. Compared to 4H-SiC, standard SiC materials exhibit greater variability in dielectric response and carrier trapping, leading to intermediate SEE sensitivity [25].

The analysis from Fig. 8.4(c) reveals that 4H-SiC exhibits a wide bandgap (~3.26 eV), high breakdown field, and strong thermal conductivity, making it exceptionally resistant to radiation-induced charge generation. According to the drift-diffusion process dominates the charge transport, but in wide-bandgap materials like 4H-SiC, fewer electron-hole pairs are generated under ionizing radiation due to the higher energy required to excite carriers. The work of Aishwarya & Lakshmi (2024) also confirms that materials with high tunneling barriers and low intrinsic carrier concentration show minimal transient peak current (I_{peak}) and collected charge (Q_C), translating into enhanced resilience under high LET radiation [20, 25].

The reduced heavy ion-induced charge density observed in SiC-4H SOI VSTB FETs (Fig. 8.5(a), compared to their Si and conventional SiC SOI counterparts, and is primarily attributed to the superior material and structural characteristics of the device. The wide bandgap of 4H-SiC results in lower ionization efficiency and a reduced intrinsic carrier concentration, which in turn limits the amount of charge generated during an ion strike.

Additionally, the high carrier recombination rate in 4H-SiC further suppresses charge collection. The SOI structure enhances radiation tolerance by providing effective electrical isolation through the buried oxide layer, thereby mitigating charge diffusion into the substrate. Furthermore, the VSTB architecture confines the electric field and reduces the sensitive volume, minimizing the overall charge collection. These combined properties make the SiC-4H SOI VSTB FET highly resilient to heavy ion irradiation, demonstrating improved single-event effect (SEE) robustness for radiation-hardened applications [21],[22]. The study further examines the transient behaviour of the device for Si, SiC and 4H-SiC while maintaining a constant LET of 20 MeV-cm^2/mg (Fig. 8.5(b). Transient analysis was conducted in the OFF state of the device, where a drain voltage of 0.5 V was applied to observe how the device responds to ion-induced disturbances. Transient drain current varies with material properties. Silicon generates more carriers due to its low bandgap and ionization energy, resulting in higher transient current. SiC and 4H-SiC, with wider bandgaps and higher ionization energies, produce fewer carriers and exhibit faster recombination, reducing charge collection. In 4H-SiC SOI VSTB structures, enhanced field confinement and substrate isolation further suppress transient response, improving radiation hardness [24-25]. Fig. 8.6 illustrates the collected charge (Q_C) over time for Si, SiC and SiC-4H at a constant LET of 20 MeV·cm²/mg. The collected charge per unit length (pC/μm) varies significantly among Si, SiC, and 4H-SiC due to differences in their fundamental material properties. Silicon exhibits the highest collected charge, attributed to its low ionization energy and long carrier lifetime, which enable efficient generation and transport of charge carriers [26-27].

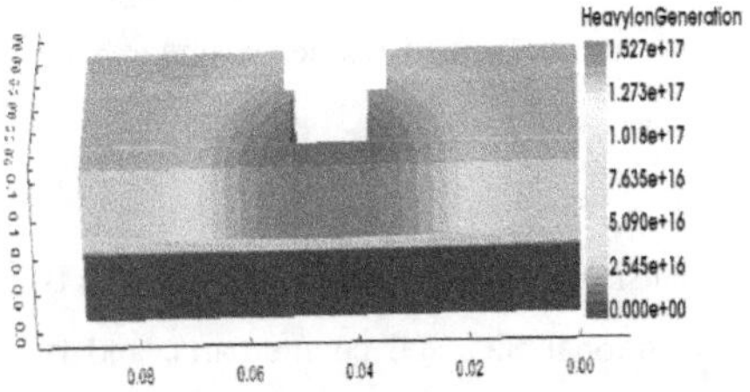

(a)

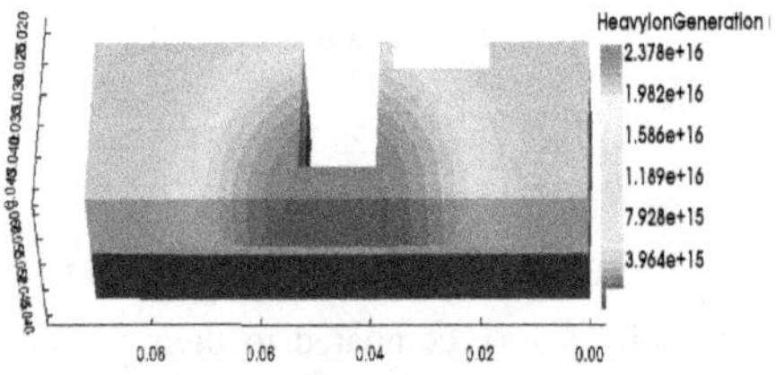

(b)

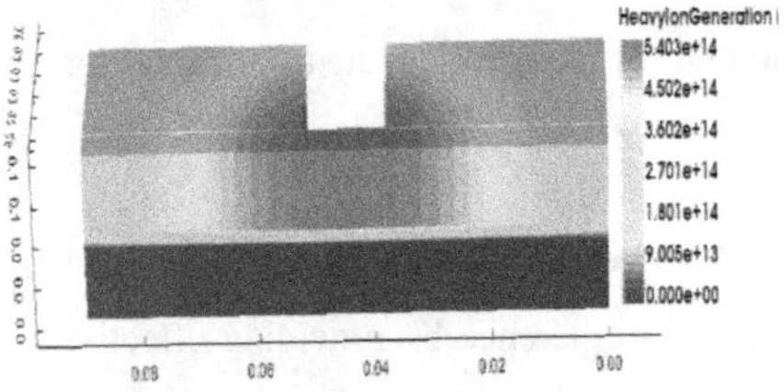

(c)

FIGURE 8.4 Heavy Ion-Generation (a) Si, (b) SiC, and (c) SiC_4H in middle (X = 44.5 nm) of the channel at Constant LET = 20 n²/mg.

In contrast, SiC collects less charge due to its higher ionization energy and faster recombination, which limit the number of carriers reaching the drain. 4H-SiC demonstrates the lowest collected charge, owing to its even higher ionization energy, enhanced recombination, and stronger electric field confinement. When combined with SOI and VSTB architectures, 4H-SiC further suppresses charge collection, making it highly suitable for radiation-hardened applications [28].

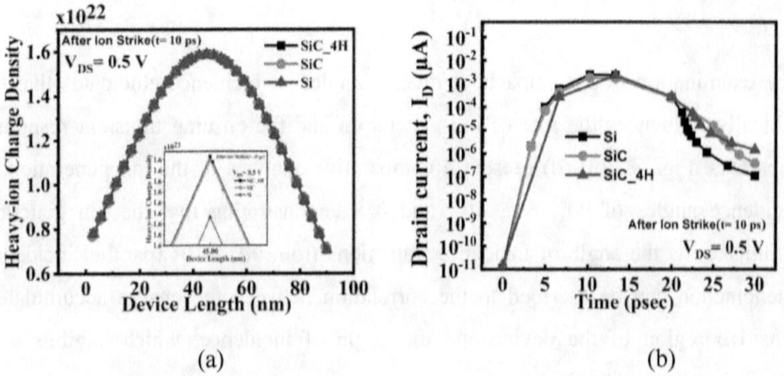

FIGURE 8.5 (a) The peak variation in Heavy-ion charge density at different positions and (b) The variation in the transient drain current with respect to the Position (x=45 nm) strikes at different LET.

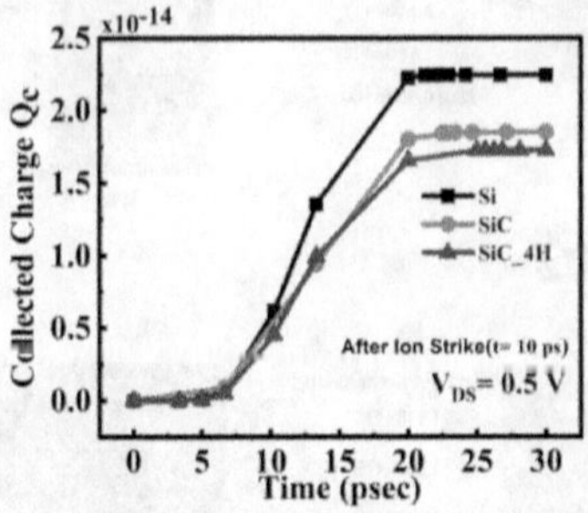

FIGURE 8.6 The collected charge (pC/μm) for Si, SiC and SiC-4H at constant LET = 20 MeV-cm^2/mg

8.4 Performance Analysis of 4H-SiC SOI VSTB FET under Varying Incidence Angles and LET Conditions

This section investigates a 4H-SiC-based SOI VSTB FET, focusing on its response to varying angles of incidence and different LET values of heavy ion and alpha radiation. The analysis aims to assess the resilience of the device to SEE and benchmark its performance against conventional Si and SiC-based devices, demonstrating the enhanced robustness and efficiency of the 4H-SiC SOI VSTB FET under radiation stress.

8.4.1 Ion Strike at different Angle of Incidence

The examination of ion impacts at diverse angles of incidence elucidates that the angle critically influences the rate of ion generation and the ensuing transient response of the apparatus. Figs. 8.7(a)-(d) present a comparative analysis of the ion generation rates for incidence angles of 90°, 60°, 30°, and 0°, demonstrating that the ion generation rate diminishes as the angle of incidence transitions from 90° to 0° (parallel incidence). This phenomenon can be ascribed to the correlation between the charge accumulated in the sensitive region of the device and the angle of incidence, which exhibits an inverse proportionality to cos (θ).

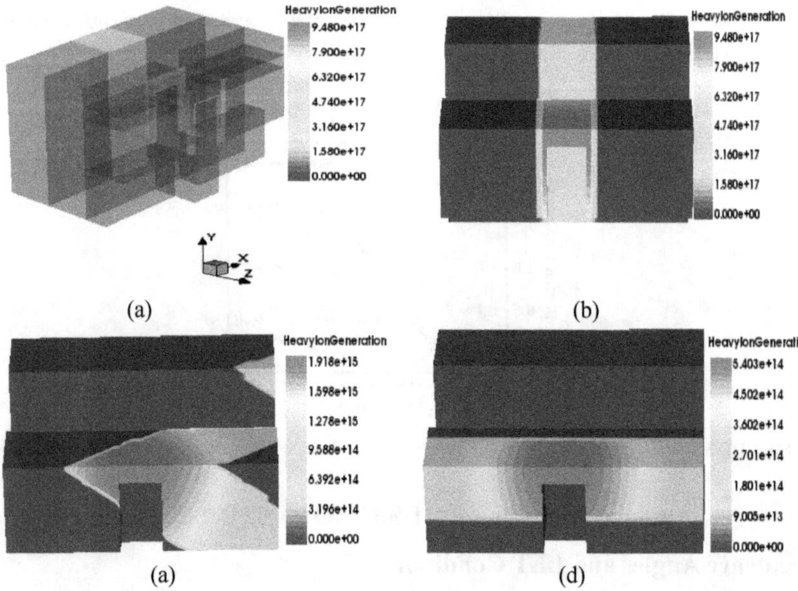

FIGURE 8.7 Heavy Ion-Generation Contour plot with respect to the different angle of incidence (a) 90^0, (b) 60^0, and (c) 0^0 at LET = 20 MeV-cm2/mg.

At augmented angles, the radiation infiltrates more profoundly into the device, thereby extending the interaction path length and amplifying the ion generation rate. As the angle of incidence escalates, the quantity of charge carriers produced concurrently increases, resulting in elevated transient currents and enhanced collected charges [29],

[30]. This trend is clearly depicted in Fig. 8.8(a) and 8.8(b), which illustrate the simultaneous increases in collected charge and drain current, respectively, as the angle of incidence becomes more acute. The extended radiation path length at higher angles fosters ionization, culminating in an augmented probability of charge generation within the sensitive volume of the device.

Furthermore, the angle of incidence is contingent upon the specific type of radiation, with heavier ions exhibiting heightened sensitivity to angular influences in contrast to lighter particles, such as alpha particles. This observation suggests that the transient response and charge collection phenomena are markedly intensified under steep-angle ion impacts, rendering normal incidence (90 degrees) the most consequential configuration. The analysis underscores the angular sensitivity of radiation effects in TFETs, thereby accentuating the imperative for angular-specific design considerations to alleviate radiation-induced susceptibilities in such devices.

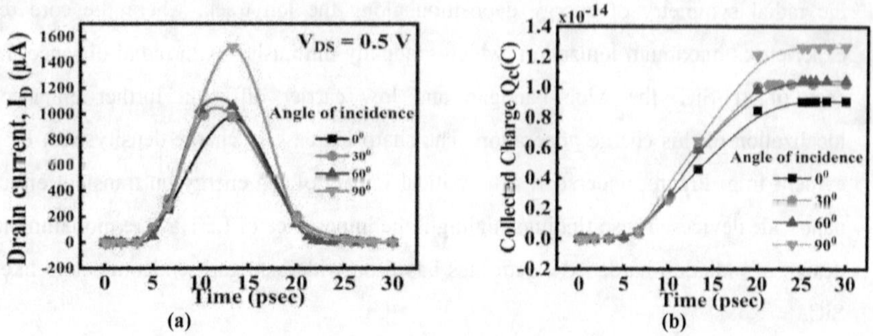

FIGURE 8.8 (a) Transient drain current and (b) Collected Charge with respect to angle of incidence at 90^0,45^0, 30^0 and 0^0 at constant LET = 20 MeV-cm2/mg

8.4.2 Sensitivity at different LET Values

Fig. 8.9 illustrates the variation in heavy-ion-induced charge density along the device length of a 4H-SiC SOI VSTB FET for different LET values ranging from 1 to 30 MeV-cm²/mg under a drain-source voltage of 0.5 V. The charge density exhibits a Gaussian distribution with the peak cantered around 45 nm, corresponding to the most radiation-sensitive region of the device.

TABLE 8.2. The comparative analysis of radiation parameters at different LET Values

LET ($MeVcm^2/mg$)	Maximum Transient Current (μA)	Deposited Charge (C/μm)	Collected Charge (C/μm)	Bipolar Gain (ß)
1	259	2.158×10^{-18}	2.070×10^{-15}	959.22
10	1410	2.237×10^{-18}	1.150×10^{-14}	5140.8
20	2297	2.224×10^{-18}	1.847×10^{-14}	8304.8
30	3085	2.245×10^{-18}	2.557×10^{-14}	11389.75

As the LET increases, the peak charge density rises significantly, reaching approximately 2.38×10^{22} cm^{-3} for LET = 30 MeV-cm^2/mg. This behaviour is attributed to the direct correlation between LET and energy deposition rate; higher LET values result in more energy being transferred to the semiconductor per unit path length, thereby generating a greater number of electron-hole pairs. The Gaussian nature of the distribution arises from the radial symmetry of energy deposition along the ion track, where the core region experiences maximum ionization, which gradually diminishes with radial distance. In the case of 4H-SiC, the wide bandgap and low carrier diffusion further enhance the localization of this charge generation. The sharp increase in charge density with LET, as evident from Figure, underscores the critical impact of ion energy on transient effects in nanoscale devices. These findings highlight the importance of LET aware modelling in the design of radiation-hardened electronics based on wide-bandgap semiconductors like 4H-SiC.

TABLE 8.3 Comparison Table of different FET under Radiation effects

S No.	Device	I_{DMax} (mA)	Max Q_C (fC)
1	LTFET[50]	260	2.1
2	DG-FeTFET [15]	272	2.1
3	Conventional TFET [52]	1600	52
4	Doping less TFET [53]	453	2.92
5	Ge-Source TFET [33]	482	34.6
6	Si –SOI VSTB FET [54]	1.481	11.82
This work	**SiC-4H-SOI VSTB FET**	**1.410**	**11.5**

The dynamics of the transient drain current I_{DS} is depicted in Fig. 8(b), are governed by the rapid separation and drift of the generated charge carriers under the influence of the electric field. Initially, I_{DS} surges within a few picoseconds due to the prompt drift of electrons and holes to respective terminals. The peak current increases significantly with LET, attaining values of 877.73 μA, 1481.02 μA, 2008.76 μA, and 2476.66 μA for LET values of 1, 10, 20, and 30, respectively, reflecting the proportionality between carrier generation rate and energy deposition. Fig. 8.9(c) further quantifies the collected charge Q_C , which represents the net charge swept to the drain terminal. The time evolution shows saturation around 20 ps due to recombination and the exhaustion of mobile carriers, with Q_C peaking near 2.1×10^{-14} C for the highest LET. Fig. 8.9(d) consolidates this relationship by plotting the maximum I_{DS} against LET, exhibiting a near-linear trend due to the enhanced carrier mobility and field-assisted charge separation at high LET.

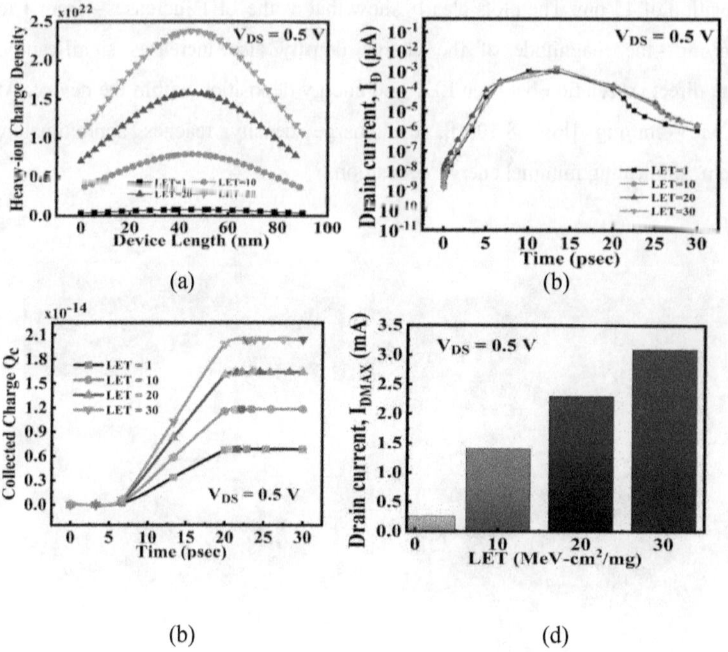

FIGURE 8.9 (a) The variation in Heavy Ion Charge density with respect to position X = 45 nm at different LETs and (b) The Variation of transient Drain current shown for different LETs with respect to time, (c) Maximum transient Drain current, and (d) Collected charge (Q_C) with respect to different LET Value.

This analysis underscores the fundamental physics linking ion energy deposition to charge generation, separation, and collection, ultimately dictating the severity of transient effects in wide-bandgap devices like 4H-SiC, which are otherwise chosen for their high radiation tolerance. Understanding these mechanisms is crucial for predicting device performance in space and high-radiation environments and for guiding the design of radiation-hardened electronics [28, 31]. Table 8.2 delineates the comparative analysis of radiation parameters across different LET values. It is observed that as the LET increases, there is a concomitant elevation in the values of transient current, collected charge, and deposited charge. This study explores the impact of heavy ion radiation on the SOI VSTB FET at a drain voltage of V_{DS}=0.5 V, focusing on a LET value of 10 MeV-cm²/mg, as shown in Table 8.3. The 4H-SiC SOI VSTB FET used in this work shows a drain current of 1481.02 μA and a collected charge of 11.82 fC. Fig 8.10(a)–(d) illustrates the spatial distribution of heavy ion-induced charge density across the 4H-SiC SOI VSTB FET for different LET values at a fixed X-position of 45 nm. The plots clearly show that as the LET increases—from 1 to 30 MeV-cm²/mg—the magnitude of the charge density also increases significantly, validating the direct correlation between LET and energy deposition within the device. At LET = 1 MeV-cm²/mg [Fig. 8.10(a)], the charge density reaches approximately 7.947×10^{20} cm^{-3} indicating minimal energy deposition.

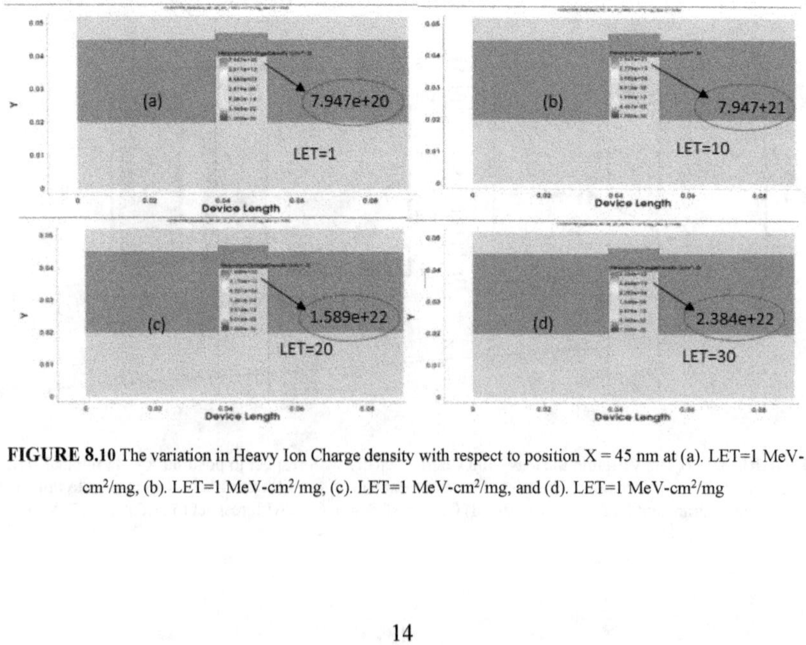

FIGURE 8.10 The variation in Heavy Ion Charge density with respect to position X = 45 nm at (a). LET=1 MeV-cm²/mg, (b). LET=1 MeV-cm²/mg, (c). LET=1 MeV-cm²/mg, and (d). LET=1 MeV-cm²/mg

As LET increases to 10 [Fig. 8.10(c)] and 20 [Fig. 8.10(b)], the peak charge density rises sharply to 1.589×10^{22} $cm^{-3}$and 7.947×10^{21} cm^{-3}, respectively. The highest charge density, 2.384×10^{22} cm^{-3}, is observed for LET = 30 MeV-cm²/mg [Fig. 8.10(d)], confirming that greater LET values result in more ionization and charge carrier generation within the silicon body. This trend is rooted in the fundamental physics of heavy ion interactions with semiconductors: as LET increases, more energy is deposited per unit path length by the incident ion, leading to increased electron-hole pair generation. This escalates the local charge density, influencing the transient response and potentially affecting the device's reliability and soft error susceptibility. Therefore, Fig. 8.10 supports the conclusion that devices exposed to higher LETs experience more intense charge build-up, which is critical for evaluating the radiation hardness of advanced semiconductor technologies.

8.4.3 Alpha Particle Generation

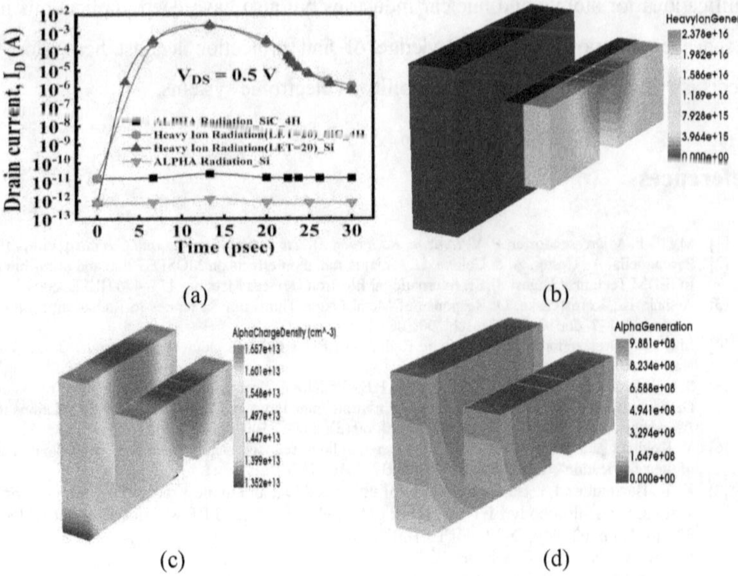

FIGURE 8.11 (a) The Variation of transient Drain current shown for Alpha particle and heavy ion radiation with respect to time (b). Alpha Particle Generation LET = 20 MeV-cm2/mg (c) Alpha charge Density and (d) Alpha generation at Constant LET = 20 MeV-cm2/mg

8.5 Conclusion

Heavy-ion radiation sensitivity of 4H-SiC SOI Vertical Super Thin Body (VSTB) FETs has been thoroughly analysed for silicon (Si), silicon carbide (SiC), and 4H-silicon carbide (4H-SiC) with 4H-SiC being reported as the most radiations hard option. Continuing with the angle of ion incidence, ɵ, VSTB FET of the 4H-SiC SOI FET was analysed for angles of 0°, 30°, 60° and 90°. It was found that increasing angle (ɵ) resulted in increased surface charge within the device. Radiation was further investigated by changing the LET values, for which a device with LET = 30 $MeV \cdot cm^2$ produced a maximum pulsed DC current of 3085 μA with a maximum collected charge of 2.557×10^{-14} C/μm. The finding of bipolar gain above 1 indicates a high level of ionic radiation sensitivity which can produce harmful effects and temporary anomalies. The energy deposition is therefore about 10^4 times higher than that produced by alpha particles since heavy-ion radiation is characterized by a much higher ionization density along its path, resulting in substantially increased damage to the 4H-SiC SOI VSTB FETs as compared to alpha particles. These findings have important ramifications for atomic and nuclear industries but also have deep implications in regards to space exploration, where knowledge of and protection against heavy-ion radiation effects is critical to guarantee the reliability of electronic systems.

References

[1] Ma, T. P. & Dressendorfer, P. V. *Ionizing Radiation Effects in MOS Devices and Circuits* (Wiley, 1989).

[2] Paccagnella, A., Cester, A. & Cellere, G. Ionizing radiation effects on MOSFET thin and ultra-thin gate oxides. In IEDM Technical Digest. IEEE International Electron Devices Meeting, 473–476 (IEEE, 2004).

[3] Arshak, K.; Korostynska, O. Response of Metal Oxide Thin Film Structures to Radiation. *Mater.Sci. Eng. B* **2006,** *133*, 1–7. doi:10.1016/j.mseb.2006.06.012

[4] Ma, T.P.; Dressendorfer, P.V. Ionizing Radiation Effects in MOS Devices and Circuits; John Wiley & Sons: New York, 1989.

[5] Schwank, J.R.; Shaneyfelt, M.R.; Dodd, P.E. Radiation Hardness Assurance Testing of Microelectronic Devices and Integrated Circuits: Radiation Environments, Physical Mechanisms, and Foundations for Hardness Assurance. *IEEE Trans. Nucl. Sci.* **Jun. 2013,** *60* (3), 2074–2100. doi:10.1109/TNS.2013.2254722.

[6] V. Koldiaev and R. Pirogova, "Vertical super-thin body semiconductor on dielectric wall devices and methods of their fabrication", U.S. Patent 8,796,085 B2, 5 Aug 2014.

[7] K. R. Barman and S. Baishya, "Study of enhanced DC and analog/radio frequency performance of a vertical super-thin body FET by high-k gate dielectrics", Int J RF Microw Comput Aided Eng., vol. 32, no. 1, pp. e22940, 2022, doi:10.1002/mmce.22940

[8] S. Roy, A. Chatterjee, D.K. Sinha, R. Pirogova, and S. Baishya, "2-D analytical modeling of surface potential and threshold voltage for vertical super-thin body FET", *IEEE Trans. Electron Devices*, vol. 64, no. 5, pp. 2106–2112, 2017,

[9] TCAD Sentaurus Device User Manual, Version J 2022.12 (Synopsis Inc.)

[10] Y. Huang et al., "Total Ionizing Dose Radiation Effects Hardening Using Back-Gate Bias in Double-SOI Structure," in IEEE Transactions on Nuclear Science, vol. 69, no. 3, pp. 453-461, March 2022, doi: 10.1109/TNS.2022.3145027.

[11] Roy Barman, K., Baishya, S. An insight to the performance of vertical super-thin body (VSTB) FET in presence of interface traps and corresponding noise and RF characteristics. Appl. Phys. A 125, 865(2019). https://doi.org/10.1007/s00339-019-3165-9

[12] R. Barman and S. Baishya, "Performance analysis of vertical super-thin body (VSTB) FET and its characteristics in presence of noise", *Appl. Phys. A*, vol. **125**, pp. 401, 2019, https://doi.org/10.1007/s00339-019-2682-x
[13] T. Skotnicki, J.A. Hutchby, T.-J. King, H.-S.P. Wong, and F. Boeuf, "The end of CMOS scaling: toward the introduction of new materials and structural changes to improve MOSFET performance", IEEE Circuits Devices Mag., vol. 21, no. 1, pp. 16–26, 2005. https://doi. org/10.1109/MCD.2005.1388765
[14] Y. Huang *et al.*, "Total Ionizing Dose Radiation Effects Hardening Using Back-Gate Bias in Double-SOI Structure," in *IEEE Transactions on Nuclear Science*, vol. 69, no. 3, pp. 453-461, March 2022, doi: 10.1109/TNS.2022.3145027
[15] F. Balestra, Silicon-on-Insulator Devices, Wiley Encyclopedia of Electrical andElectronics Engineering,edited by John G. Webster (Wiley, 2014).
[16] Li, Y. (2011). A Review of 4H-SiC Power Electronic Devices. Research & Progress of SSE. https://en.cnki.com.cn/Article_en/CJFDTOTAL-GTDZ201103004.htm
[17] Chaudhary, Rashi & Saha, Rajesh. (2023). Physical Insights of Interface Traps and Self-heating effect on electrical response of DMG FinFETs in Overlap and Underlap configurations: Analog/RF perspective. Physica Scripta. 99. 10.1088/1402-4896/ad16b0
[18] S. Tiwari, R. Saha, G. Gopal and T. Varma, "Sensitivity Analysis of Ge Source Double Gate TFET Under Heavy Ion Irradiation," in IEEE Sensors Journal, vol. 24, no. 22, pp. 36730-36736, 15 Nov.15, 2024, doi: 10.1109/JSEN.2024.3463870.
[19] Salemi, A. (2017). Silicon Carbide Technology for High- and Ultra-High-Voltage Bipolar Junction Transistors and PiN Diodes. http://www.diva-portal.org/smash/record.jsf?pid=diva2:1054818
[20] Bellone, S., Di Benedetto, L., & Licciardo, G. D. (2012). A Quasi-One-Dimensional Model of the Potential Barrier and Carrier Density in the Channel of Si and 4H-SiC BSITs. *IEEE Transactions on Electron Devices, 59*(9), 2546–2549. https://doi.org/10.1109/TED.2012.2203601
[21] Yuan, S., [30] Zhang, Q., Liu, Y., & Zhang, Y. (2012). 4H-SiC and Si Based IGBTs/VDMOSFETs Characters Comparison. Symposium on Photonics and Optoelectronics, 1–4. https://doi.org/10.1109/SOPO.2012.6271084
[22] M. Gerold, G. Dollinger, L. Günther, J. Reindl, M. Rüb,Sensitivity study of super-junction power MOSFETs by spatial and depth resolved heavy ion SEE mapping with various bias, LETs and ion ranges,Microelectronics Reliability,Volume 155,2024,115309,ISSN 0026-2714, https://doi.org/10.1016/j.microrel.2023.115309
[23] C. Peng et al., "Influence of the Total Ionizing Dose Irradiation on 130 nm Floating-body PDSOI NMOSFETs," in IEEE Transactions on Nuclear Science, vol. 62, no. 1, pp. 314-322, Feb. 2015, doi: 10.1109/TNS.2015.2390636.
[24] C. -H. Yu, Y. Wang, M. -T. Bao, X. -J. Li, J. -Q. Yang and F. Cao, "Impact of Heavy-Ion Irradiation in an 80-V Radiation-Hardened Split-Gate Trench Power UMOSFET," in IEEE Transactions on Electron Devices, vol. 69, no. 2, pp. 664-668, Feb. 2022, doi: 10.1109/TED.2021.3135369.
[25] K. Aishwarya, and B. Lakshmi. "TCAD simulation study of heavy ion radiation effects on hetero junctionless tunnel field effect transistor." Scientific Reports, vol. 14, no. 1, pp. 7643, 2024, doi. 10.1038/s41598-024-58371-6
[26] C. Spoorthi, J. Grace, A. Chavan and C. S. M. Rao, "Total Ionizing Dose and Radiation Particle Strike Analysis of Nanoscale FinFET Devices and Circuits," 2020 4th International Conference on Electronics, Communication and Aerospace Technology (ICECA), Coimbatore, India, 2020, pp. 479-484, doi: 10.1109/ICECA49313.2020.9297573.
[27] Stapor, W. J., McDonald, P. T., Knudson, A. R., Campbell, A. B., & Glagola, B. G. (1988). Charge collection in silicon for ions of different energy but same linear energy transfer (LET). IEEE Transactions on Nuclear Science, 35(6), 1585–1590. https://doi.org/10.1109/23.25502
[28] Blažeka, D. (2017). Transportna svojstva poluvodičkih detektora. https://zir.nsk.hr/islandora/object/pmf:1620/datastream/PDF/download
[29] Makino, T., Deki, M., Iwamoto, N., Onoda, S., Hoshino, N., Tsuchida, H., Hirao, T., & Ohshima, T. (2013). Heavy-Ion Induced Anomalous Charge Collection From 4H-SiC Schottky Barrier Diodes. IEEE Transactions on Nuclear Science, 60(4), 2647–2650. https://doi.org/10.1109/TNS.2013.2243469
[30] N. Kamal, M. Panchore and J. Singh, "3-D Simulation of Junction- and Doping-Free Field-Effect Transistor Under Heavy Ion Irradiation," in IEEE Transactions on Device and Materials Reliability, vol. 18, no. 2, pp. 173-179, June 2018, doi: 10.1109/TDMR.2018.2811493
[31] Y. Sun, J. Shao, Z. Liu, X. Li, Y. Liu, and Y. Shi, ''Evaluation of single event-transient effects in reconfigurable field effect transistor beyond 3 nm technology node,'' IEEE Trans. Electron Devices, vol. 68, no. 12, pp. 6001–6006, Dec. 2021, doi: 10.1109/TED.2021.3119006
[32] B. Ye, L.-H. Mo, P.-F. Zhai, L. Cai, T. Liu, Y.-N. Yin, Y.-M. Sun, and J. Liu, ''Impact of heavy ion energy and species on single-event upset in commercial floating gate cells,'' Microelectron. Rel., vol. 120, May 2021, Art. no. 114128, doi: 10.1016/j.microrel.2021.114128.
[33] H. Liu, M. Cotter, S. Datta, and V. Narayanan, ''Soft-error performance evaluation on emerging low power devices,'' IEEE Trans. Device Mater. Rel., vol. 14, no. 2, pp. 732–741, Jun. 2014, doi: 10.1109/TDMR.2014. 2316505
[34] Colinge, Jean-Pierre & Lee, C & Afzalian, Aryan & Dehdashti Akhavan, Nima & Yan, Ran & Razavi, Pedram & O'Neill, Brendan & Blake, Alan & White, Mary & Kelleher, Anne-Marie & Mccarthy, Brendan & Murphy, Richard, "Nanowire transistors without junctions", Nature nanotechnology, 2010, 5. 225-9. 10.1038/nnano.2010.15.

[35] Q. Wang, H. Liu, S. Wang, and S. Chen, ''TCAD simulation of single event-transient effects in L-shaped channel tunneling field-effect transistors,'' IEEE Trans. Nucl. Sci., vol. 65, no. 8, pp. 2250–2259, Aug. 2018, doi: 10.1109/TNS.2018.2851366
[36] G. Tian, J. Bi, G. Xu, K. Xi, X. Yang, H. Yin, Q. Xu, and W. Wang, ''Heavy ion induced single-event-transient effects in nanoscale ferroelectric vertical tunneling transistors by TCAD simulation,''Semicond. Sci. Technol., vol. 35, no. 10, Oct. 2020, Art. no. 105010, doi: 10.1088/1361- 6641/aba549.
[37] Y. Wu and Y. Takahash, '' The impact of tunnel FET on heavy ion induced transient effect,'' in Proc. ECS Meeting. Chesapeake, VA, USA, Oct. 2017, doi: 10.1149/MA2017-02/26/1143
[38] M. Sharma, R. Narang, M. Saxena, and M. Gupta, ''Single event transient effect on tapered angle hetero-junction dopingless TFET for radiation sensitive applications,'' in Proc. IEEE VLSI Device Circuit Syst. (VLSI DCS), Feb. 2022, pp. 261–264, doi: 10.1109/VLSIDCS53788.2022.9811459.

CHAPTER 9

Conclusion and Scope for Future Work

9.1 Summary of the Work

This thesis presented a detailed exploration of the SOI VSTB FET, focusing on its potential as a next-generation transistor architecture for overcoming the limitations of conventional bulk MOSFETs and even modern FinFETs. Throughout the thesis, a systematic investigation of the SOI VSTB FET was conducted across various dimensions—electrical, RF/analog, thermal, material, and radiation performance through calibrated TCAD simulations.

The study began by comparing the electrical characteristics of SOI VSTB FET with its bulk counterpart. It was found that the presence of a buried oxide (BOX) in SOI structures dramatically reduced short-channel effects (SCEs) by isolating the active channel from the substrate, suppressing parasitic charge sharing between source and drain. The SOI VSTB FET consistently demonstrated lower off-state leakage, a higher ON-state current, and a reduced sub-threshold swing (SS) compared to bulk VSTB devices, establishing it as a power-efficient and reliable architecture for future CMOS technology.

Next, the RF and analog performance of SOI VSTB FETs was analyzed in depth. Key performance indicators such as transconductance (g_m), cut-off frequency (f_T), and gain-bandwidth product (GBP) were all found to improve substantially in the SOI-based architecture. Additionally, the device showed enhanced noise immunity, particularly under conditions of low interfacial trap densities, making it suitable for high-speed and mixed-signal circuit applications.

Further, investigations revealed that self-heating effects (SHE) remain a significant challenge for ultra-scaled SOI VSTB FETs, especially under high drive current conditions. The results showed a clear degradation in both electrical and RF parameters at elevated temperatures, emphasizing the need for improved thermal management strategies and device-level optimizations to address heating concerns. Nonetheless, the SOI architecture, aided by STI-supported fins, still provided better thermal handling compared to conventional planar MOS structures.

To further enhance carrier transport, channel material engineering was explored using $Si_{1-x}Ge_x$ heterojunctions. Results demonstrated that increasing Ge content improved electron mobility and ON-state current, albeit at the cost of slightly higher leakage currents and degraded SS. The trade-off was manageable with careful optimization of oxide thickness and channel dimensions, making SiGe-channel SOI VSTB FETs particularly attractive for high-performance, low-voltage applications.

Additionally, a comparative analysis between Si and Ge SOI VSTB FETs revealed that while Ge devices provide superior drive current, they suffer from higher leakage and degraded subthreshold characteristics. However, applying architectural stress further improved the electrical properties of both Si and Ge channels, emphasizing the importance of strain engineering in nanoscale device optimization.

A critical contribution of this work was the radiation resilience study, essential for applications in space and defense. The analysis showed that 4H-SiC SOI VSTB FETs outperformed their Si and SiC counterparts, offering superior radiation hardness, reduced collected charge under heavy-ion strikes, and better tolerance to single-event transients (SETs). These results confirmed the applicability of SOI VSTB FETs for radiation-hardened circuits, expanding their relevance beyond conventional consumer electronics.

In conclusion, this thesis provides comprehensive validation of SOI VSTB FETs as a promising candidate for advanced CMOS nodes, successfully addressing challenges related to SCEs, power efficiency, RF performance, noise resilience, thermal effects, and radiation hardness. The extensive insights gained in this work not only demonstrate the viability of SOI VSTB FETs for high-performance (HP) and low-power (LP) applications but also establish a strong foundation for future exploration in both academic research and industrial deployment.

9.2 Further Scope of Work

As the SOI VSTB FET is still a relatively new entrant in the family of advanced semiconductor devices [1-2], it offers tremendous potential for future research and technological development. Its 3D vertical structure combined with SOI technology provides a unique platform [3-4] to overcome the limitations of conventional FinFETs and other planar devices, making it highly suitable for next-generation electronics. While this thesis has explored its electrical, RF/analog, linearity, radiations, and thermal behavior [5-6] in detail, much remains to be discovered to fully realize its potential.

One of the most crucial directions for future research is compact modelling of SOI VSTB FETs [7-8]. Developing accurate and efficient analytical or empirical models will be essential for integrating these devices into circuit-level simulations [9-10], enabling practical use in logic, memory, and RF circuit designs. Alongside modeling, thermal management strategies [11-12] need to be advanced, as self-heating effects [13] (SHE) become significant at nanoscales due to the insulating nature of buried oxide layers. Incorporating advanced heat spreading techniques and optimizing STI geometry could help mitigate these issues. Another important avenue is material engineering [14]. Exploring III-V semiconductors such as InGaAs or GaN [15-16] as channel materials can significantly improve carrier mobility and ON-state current, further enhancing device performance. Additionally, tailoring the buried oxide properties and interface characteristics [17] may unlock new design optimizations for low-power and high-speed applications. In parallel, radiation hardness [18-19] needs to be systematically studied, especially for space and defense applications. Future work should focus on evaluating Total Ionizing Dose (TID) effects [20], charge trapping phenomena, and Single Event

Upsets (SEU) in the SOI VSTB architecture. Experimental validation using heavy-ion, proton, and gamma-ray [21-22] exposure is necessary to certify its reliability in harsh radiation environments like satellites and deep-space missions. SOI VSTB FETs can emerge as a key enabler for future ultra-low-power, high-performance, and radiation-hardened semiconductor technologies.

References

[1] S. Roy, A. Chatterjee, D. K. Sinha, R. Pirogova, and S. Baishya, "2-D Analytical Modeling of Surface Potential and Threshold Voltage for Vertical Super-Thin Body FET," IEEE Trans. Electron Devices, vol. 64, no. 5, pp. 2106–2112, May 2017, doi: 10.1109/TED.2017.2687465.

[2] K. R. Barman and S. Baishya, "A compact model for surface potential and subthreshold slope for vertical super-thin body (VSTB) FET," Appl. Phys. A, vol. 125, no. 401, 2019. https://doi.org/10.1007/s00339-019-2700-z

[3] J.-P. Colinge et al., "Nanowire transistors without junctions," Nature Nanotechnology, vol. 5, pp. 225–229, 2010. https://doi.org/10.1038/nnano.2010.15

[4] I. Ferain, C. A. Colinge, and J.-P. Colinge, "Multigate transistors as the future of classical metal–oxide–semiconductor field-effect transistors," Nature, vol. 479, pp. 310–316, 2011. https://doi.org/10.1038/nature10676

[5] J.-P. Raskin, T. M. Chung, V. Kilchytska, D. Lederer, and D. Flandre, "Analog/RF performance of multiple gate SOI devices: Wideband simulations and characterization," IEEE Trans. Electron Devices, vol. 53, no. 5, pp. 1088–1095, May 2006, doi: 10.1109/TED.2006.871876.

[6] D. Esseni and P. Palestri, Device Modeling for Analog and RF CMOS Circuit Design, Cambridge University Press, 2011.

[7] M. J. Kumar and A. Chaudhry, "Two-dimensional analytical modeling of fully depleted SOI MOSFET and its application for the design of novel circuits," IEEE Trans. Electron Devices, vol. 49, no. 3, pp. 481–489, Mar. 2002, doi: 10.1109/16.989429.

[8] A. Kranti and G. A. Armstrong, "Nonclassical Channel Design in MOSFETs for Improving OTA Gain-Bandwidth Trade-Off," IEEE Trans. Circuits Syst. I, vol. 57, no. 12, pp. 3048–3054, Dec. 2010, doi: 10.1109/TCSI.2010.2071470.

[9] B. Yu, L. Chang, S. Ahmed, H. Wang, S. Bell, C. Tabery, et al., "FinFET scaling to 10 nm gate length," IEDM Tech. Dig., 2002, pp. 251–254, doi: 10.1109/IEDM.2002.1175828.

[10] Y. Taur and T. H. Ning, Fundamentals of Modern VLSI Devices, 2nd ed., Cambridge Univ. Press, 2009.

[11] H. Wang, Y. Taur, and R. Huang, "Self-Heating Effects of SOI and Bulk FinFETs," IEEE Electron Device Lett., vol. 32, no. 1, pp. 44–46, Jan. 2011, doi: 10.1109/LED.2010.2087785.

[12] D. J. Frank et al., "Design and Performance Considerations for Nanoscale MOSFETs," IBM J. Res. Dev., vol. 46, no. 2–3, pp. 265–286, Mar.–May 2002.

[13] S. E. Liu et al., "Self-heating effect in FinFETs and its impact on devices reliability characterization," IEEE Int. Rel. Phys. Symp., 2014, pp. 4A.4.1–4A.4.4, doi: 10.1109/IRPS.2014.6860642.

[14] A. M. Ionescu and H. Riel, "Tunnel field-effect transistors as energy-efficient electronic switches," Nature, vol. 479, pp. 329–337, 2011. https://doi.org/10.1038/nature10679

[15] M. M. Waldrop, "More than Moore," Nature, vol. 530, pp. 144–147, Feb. 2016. https://doi.org/10.1038/530144a

[16] R. Chau et al., "High-k/metal-gate stack and its MOSFET characteristics," IEEE Electron Device Lett., vol. 25, no. 6, pp. 408–410, Jun. 2004, doi: 10.1109/LED.2004.828876.

[17] N. Miyata, "Study of direct-contact HfO2/Si interfaces," Materials, vol. 5, no. 3, pp. 512–527, 2012, https://doi.org/10.3390/ma5030512.

[18] M. J. Gadlage, R. D. Schrimpf, R. A. Reed, et al., "Single Event Upset and Charge Collection in FinFETs and SOI Devices," IEEE Trans. Nucl. Sci., vol. 53, no. 6, pp. 3245–3250, 2006, doi: 10.1109/TNS.2006.887933.

[19] S. Baishya, "Radiation Effects on Advanced Semiconductor Devices: A Review," Microelectronics Reliability, vol. 61, pp. 1–7, 2016. doi: 10.1016/j.microrel.2016.03.009.

[20] D. Fleetwood, "Total ionizing dose effects in MOS and low-dose-rate-sensitive linear-bipolar devices," IEEE Trans. Nucl. Sci., vol. 60, no. 3, pp. 1706–1730, Jun. 2013, doi: 10.1109/TNS.2013.2243756.
[21] A. Holmes-Siedle and L. Adams, Handbook of Radiation Effects, Oxford Univ. Press, 2002.
[22] M. Gasiorek, R. Gaillard, and R. Velazco, "Heavy ion testing of nanoscale SRAM devices," IEEE Trans. Nucl. Sci., vol. 60, no. 6, pp. 4580–4587, Dec. 2013.